IDEAS, INVESTIGATION and THOUGHT

Sibilla E. Kennedy

Polytechnic Institute of New York
and Hofstra University

Esther B. Sparberg
Frances S.K. Sterrett
Roselin S. Wagner

Department of Chemistry
Hofstra University

AVERY PUBLISHING GROUP INC.
Wayne, New Jersey

Copyright © 1978, 1980 by Sibilla E. Kennedy, Esther B. Sparberg, Frances S.K. Sterrett, and Roselin S. Wagner.

ISBN 0-89529-126-6

All rights reserved. No part of this publication may be reproduced, stored in a retrieval system, or transmitted, in any form or by any means, electronic, mechanical, photocopying, recording, or otherwise, without the prior written permission of the author.

Printed in the United States of America

CONTENTS

PREFACE

Although various laboratory manuals have been published recently that emphasize an open-ended "problem" approach, they are suitable only for small laboratory classes. We strongly felt the need for a laboratory manual that could be used in classes as large as thirty students, and yet that would be stimulating, provocative, and that would give insight into the meaning of ideas in chemistry.

The goals of our manual, and method of carrying out these goals, are as follows:

1. We have attempted to prepare the student for an understanding of the concepts involved in a particular experiment by prefacing each experiment with a series of leading questions that he must answer prior to carrying out the experiment. These questions are assigned the week before the experiment is to be done. They are handed in at the beginning of the first laboratory period for which the experiment has been assigned. Although the student is therefore made aware of the idea involved, the preliminary questions are carefully formulated so that they avoid giving the student too much information on the problem he is preparing to investigate.

2. Each experiment is preceded by a brief historical discussion of the evolution of the ideas relating to the experiment. This analysis of how the concepts developed, and what individuals contributed to their development, emphasizes chemistry as an evolving, dynamic, changing field of knowledge which owes its growth to the efforts of many people from many countries. The usual authoritarian approach of teaching chemistry as a "crystallized, authoritative, anonymous" body of knowledge, as Bentley Glass wrote about science in general, is so demolished. Chemistry in the laboratory becomes a study of ideas and their genesis in the physical world. This structure of ideas is seen as the product of the minds of individuals, rather than as an area of knowledge completely cut off from its relationship to human beings.

3. The experiments are chosen so that they illustrate the significant concepts in chemistry. They include experiments in phsyical chemistry (e.g., solid-state problems, chemical equilibria, and reaction rates), quantitative analysis, semimicro qualitative analysis, and inorganic preparations which emphasize some specialized techniques. To stress the "problem" approach, most experiments include the use of "unknowns" and quantitative determinations.

The order of experiments is mostly sequential because we feel that this illustrates the cumulative nature of science (although some experiments may be omitted without losing the trend of the developing ideas). This approach has the additional advantage of stimulating the student to use ideas, concepts, information, and techniques of previous experiments, instead of routinely following instructions which have little relation to previous laboratory experiences.

Students are required to write balanced equations for all reactions.

4. The experiments are followed by a series of concluding questions (Thought) the student answers after he has completed each experiment. Their purpose is to further aid the student in extracting from the apparent chaos of the laboratory experience the most significant ideas that can be inferred by carrying out a particular experiment. The questions emphasize these points: What inference is one justified in making based on the results of the experiment? Can a valid hypothesis be formulated on the basis of the experiment? What are the limitations of the particular experiment? How precise are the measurements? What is the relationship between this experiment and various concepts accepted in science?

The preliminary questions, the historical discussion, the experiment itself, and the questions at the end of each experiment point up that fruitful concepts and theories in chemistry are based upon a solid foundation of study of the physical world. Too often, chemistry courses introduce concepts that are never justified to the student. In the laboratory there exists a superb opportunity to illustrate the interplay between theory and experimentation. The aim of this manual is to emphasize this relationship.

SAFETY RULES

1. *Wear safety glasses or prescription glasses at all times while you are in the laboratory.* Protect your clothing with a lab coat or apron.

2. Work in the laboratory *only with supervision.*

3. Perform unassigned or unlisted experiments *only with written approval* of your instructor.

4. Never remove chemicals or supplies from the laboratory.

5. Report all accidents and injuries, however minor, to the instructor.

6. If any liquid reagents are spilled, *flood the area with water* immediately.

   a. For acid burns: Saturated sodium hydrogen carbonate ($NaHCO_3$) solution can also be used.

   b. For alkali burns: Saturated boric acid (H_3BO_3) solution can also be used.

7. For inserting glass tubing into a rubber stopper:

   a. The glass tubing must be fire-polished.

   b. Wrap your hands in a towel, and hold the glass near the stopper.

   c. Do not force the glass into the hole. Use a gentle twisting motion, and a lubricant, for example, water or glycerine.

8. *Do not taste any chemicals* unless so directed. Should you accidentally imbibe any reagent, rinse your mouth with water and report to the instructor immediately.

9. When examining the odor of a reagent or reaction, do not put your head directly over the container. Use your hand to fan the odors in the direction of your nose and inhale carefully.

10. If poisonous or noxious gases are going to be evolved from your experiment, perform the experiment in the hood.

11. When diluting concentrated acids, be sure to *add the acid* to water, while stirring the mixture.

12. For burns caused by heat, apply ice or cold water immediately.

13. For disposal of waste material:

    a. Put all paper, matches, and broken glass into the wastebasket.

    b. Put soluble solids in the sink, dissolve them with a large quantity of water from the tap, and wash them down the drain.

    c. Pour acids, bases, and other liquids down the drain with much water.

    d. Put insoluble solids in specially assigned jars.

    e. For special waste problems, your instructor will give directions.

14. For reagents:

    a. *Do not* remove reagent bottles from the shelf *at any time*.

    b. To obtain chemicals, take a watch glass or beaker to the reagent shelf with you. Using the techniques described on pages 7–9, remove from the reagent bottle only what you need.

    c. *Do not* place corks or stoppers down on the laboratory bench. Hold the corks or stoppers in the hand as shown by the instructor and as pictured on page 8.

    d. *Do not* return reagents to their bottles *at any time*.

    e. *Do not* insert pipets or droppers into reagents.

    f. Read all labels twice, once before and once after removing the chemical.

    g. When you finish using the reagent, be sure to close the bottle with its own cork, cover, or stopper. Reagents in bottles *must not* be left standing uncovered.

15. For glassware:

    a. All glass tubing *must be fire-polished* before use.

    b. Do not heat graduated cylinders or glass bottles *at any time*.

16. Always use a glass or porcelain container for weighing any substance on the balance.

17. Keep books and papers away from working areas to avoid their being damaged.

18. Anticipate danger. Know the location of the fire extinguisher and the emergency shower.

19. Before leaving the laboratory, *wash your desk and wipe it dry*.

EXPERIMENTAL LIMITATIONS

ACCURACY AND PRECISION

In the study of science it is important to have an exact understanding of the meaning and use of words in that science. The definitions of words in science are specific. This is true of the words "accuracy" and "precision."

The word "accuracy" is used to describe the closeness that a measurement has to a standard. The standard may have an arbitrary value, as does the value of the atomic weight of carbon (12). If the value of the standard is not arbitrary, it is derived from empirical information obtained from many experiments and is an accepted value. This value, unlike the value for carbon (12), which cannot change unless the definition of the standard changes, may change as more experimental values accumulate. The relative atomic weights of all the other elements are examples of accepted values.

The word "precision" is used to evaluate measurement. The precision of a particular set of measurements depends upon the measuring instrument and the degree of variation within the series of values. These values are obtained operationally by the actual act of measuring several times in the same way with the same instruments. For example, a notebook page is measured with a meter stick marked off in tenths of centimeter units and found to be 23.7 centimeters long. Repeated measurements of the same page reveal that the estimate may vary by ± 0.1 cm. The precision of these measurements with this ruler is limited by this amount.

Values in very close agreement, however, do not necessarily imply greater accuracy or more reliable values. The calibration of the ruler may be in error and the ruler may not be a good copy of the accepted standard.

Suppose that by arithmetic manipulation, as for example finding the mean of a set of values, another digit is obtained, the number 27.71 instead of 27.7. This does not increase the accuracy of the measurement. As a matter of fact, reporting the value to the hundredth place in this case is misleading, because it implies a known value in the hundredth place and consequently gives it physical meaning. (The measurement, however, *appears* to be more accurate.)

There are several different ways data can be evaluated. For example, the precision of a set of measurements is often reported as the average deviation from the mean value of a series of measurements. A book page, for example, is measured four times and the following values are found: 27.6, 27.5, 27.7, and 27.5 cm. The mean (or average) value is 27.57, or 27.6 cm. The deviation is the difference between the experimental value and the mean.

Expt. value (cm)	Mean (cm)	Deviation from mean	Average deviation
27.6		0	0.08
27.5	27.6	– 0.1	
27.7		0.1	
27.5		– 0.1	

Average deviation is the mean value for all the deviations taken without regard to sign. To calculate the relative average deviation, or percent (parts per hundred) error, divide the average deviation by the mean value of the experimental values and multiply by 100.

$$\frac{0.08 \text{ cm}}{27.6 \text{ cm}} \times 100 = 0.3\% \text{ error}$$

A second method for indicating the measure of precision is the range, which is the difference between highest and lowest values. For the above example this is 0.2.

A third method of evaluating data, computation of standard deviation, gives a more precise picture of the value of the data when a large sample of information is presented. The equation for the determination of standard deviation (s) is:

$$S = \sqrt{\frac{\Sigma(\Delta)^2}{n-1}}$$

Δ = the deviation from the mean.

n = the number of samples.

Example

The percent of magnesium in magnesium oxide was determined by a class of students. The results of the class, the deviations (Δ) and $(\Delta)^2$ are recorded in the following table.

Experiment Number	% Mg	Δ	$(\Delta)^2$
1	50	–12	144
2	59	–3	9
3	63	+1	1
4	71	+9	81
5	58	–4	16
6	68	+6	36
7	59	–3	9
8	59	–3	9
9	59	–3	9
10	62	0	0
11	72	+10	100
12	67	+5	25

$\text{Mean} = \frac{747}{12} = 62$ $\quad \Sigma(\Delta)^2 = 439$ $\quad n - 1 = 11$

$$S = \sqrt{\frac{439}{11}} = 6.3$$

The results for this class's data would be reported as

$$\% \text{ Mg} = 62 \pm 6.3$$

Such a large standard deviation suggests there may be one or more uncontrolled variables.

Quite often, as in the set of data above, there may be values which do not seem to fit in. The question arises, therefore, whether any of this data can be deleted. The only sure answer is to reject any data which are known to be faulty because of the way in which the experiment was conducted. For anything else the answer is extremely difficult.

There are, however, several accepted statistical techniques. One method is to remove temporarily a questionable value and then recalculate a new mean and a new standard deviation. If the temporarily discarded value is more than four standard deviations away from the new mean value, it can probably be discarded.

To illustrate, in the above sample assume that 50 is the datum to be rejected. Recalculation of the mean gives 63 and of the standard deviation, 5.2. Multiplying: 5.2 x 4 = 21, and then subtracting this number from the mean (63 - 21) gives 42. Since the number 50 lies inside the range of acceptable data it can not be rejected. A similar test for 72 % Mg reveals that this value also can not be rejected.

To sum up, precise measurements depend upon the reliability of the instrument used and the agreement of the values for several careful measurements. Accuracy refers to the degree to which measurements approach a most probable or accepted value and is limited by the sum of the errors inherent in the measurements (absolute error or relative error). These kinds of errors are discussed on the following pages. For a more complete discussion of this aspect, see the section on significant figures.

ABSOLUTE AND RELATIVE ERROR AND PERCENT ERROR

Absolute error can be defined as the *difference* between experimentally measured values and the accepted measured value. To refer to a previous example, if the accepted value for the length of the book page is 27.6 cm and the average value of your measurements is 27.7 cm, the absolute error is 0.1 cm.

Relative error may be expressed in terms of percent or parts per thousand. In the example previously given, the relative error of the measurement of the length of a book page would be calculated as:

$$\frac{0.1}{27.6} \times 100 = 0.4\%$$

or 4 parts per thousand.

Often no accepted value is known. Then, error is based only on experimental values. It is defined as the average deviation from the mean value of the experimental data, as described in the previous section on precision.

It can be reasoned from the foregoing that the type of instrument used for any particular measurement has its own limiting precision and accuracy. The choice of instrument to be used is based upon the degree of accuracy required under a given set of circumstances. For example, it is not necessary to know to the pint what the capacity of your car's gasoline tank is because the chances are the fuel pump cannot make use of that last pint. So a device that reads E when you are on that last gallon is a sufficiently accurate measurement. On the other hand, you may want to know to the milligram just how much aspirin you are taking. The following is a list of measuring devices found in your laboratory drawer and the usual precision of measurement for each one. Study the table carefully so that you will be aware of the degree of precision you can expect when you perform your experiments. In addition, many instruments are marked with the manufacturer's calibration and precision evaluation.

You can expect error in measurement from two sources. One comes from the limiting precision of your instruments, that is, the sensitivity of the instruments. The second source of error comes from inaccurately calibrated or malfunctioning instruments, which consequently give incorrect values. The second source can give precise but inaccurate results. (Negligent errors do not fall into this category because they result from careless or improper use of instruments.) However, it might be noted here that no measurements, regardless how precise and accurate the instrument or how well-trained and careful the operator, can ever be known with absolute certainty. This is due partially to the many physical variables which cannot be controlled. These are sometimes referred to as random or indeterminate errors.

Instrument	Precision of measured value	Range
Top-loading centigram balance	± 0.01 g	0.02 g
Analytical balance, single pan	± 0.0001 g	0.0002 g
50-ml buret	± 0.02 ml	0.04 ml
25-ml transfer pipet	± 0.02 ml	0.04 ml
5-ml transfer pipet	± 0.01 ml	0.02 ml
250-ml volumetric flask	± 0.10 ml	0.20 ml
50-ml graduated cylinder	± 0.3 ml	0.6 ml
10-ml graduated cylinder	± 0.2 ml	0.4 ml
110°C thermometer	± 0.3°C	0.6°C

In the past, careful attention to error has often led to discoveries. The discovery of argon in the atmosphere resulted from the careful investigations by Lord Rayleigh and Sir William Ramsay of puzzling findings on the density of nitrogen. Rayleigh had been probing the weight relationships between hydrogen and oxygen to find out whether oxygen and other elements also were indeed multiples of the atomic weight of hydrogen, as had been suggested by W. Prout. During the course of his work, he found that nitrogen obtained from ammonia was slightly lighter than nitrogen obtained from the atmosphere, by about 5 parts in 1000. This small difference was significant to the investigator because he knew the error that could be expected from the experimental setup, and the values obtained exceeded the limits of the experimental error. Therefore, instead of attributing the puzzling findings to instrumental error, many repetitions of the experiment were carried out. These determinations revealed that there was always the same difference in the density of nitrogen prepared by the different methods. Interestingly, an additional clue that he was on the trail of something new was obtained by similar experiments to prepare oxygen by different methods. Oxygen, unlike nitrogen, showed a consistent density regardless of method of preparation. After various hypotheses on the anomalous findings on nitrogen were suggested and discarded, an ingenious set of experiments completed in 1894 by Ramsay revealed the presence in atmospheric nitrogen of very small quantities of a new gas denser than nitrogen. It was named argon, "the lazy one."

The careful attention paid to minute, quantitative details has frequently paid enormous dividends. Lord Kelvin expressed this in the vivid quotation which concluded: "But nearly all the grandest discoveries of science have been but the rewards of accurate measurement and patient long-continued labor in the minute sifting of numerical results."

SIGNIFICANT FIGURES

A necessary corollary to the concept of precise measurement and accurate results is the concept of significant figures. Only those figures should be reported which you know to be reliable values. For example, if a stray dog comes to live at your house, you can estimate from his size and type that he may be 1 year old. But you certainly could not say with any degree of certainty that the dog is 11 months, 3 days old. This statement would probably be false. If you weigh a sample on a centigram balance, the weight can be reported as 25.50 grams, but not as 25.5032 g. Significant figures of any measurement are those known with certainty plus the last number, which is always an estimate. The weight of the sample on a centigram balance may vary with succeeding weighings as follows: 25.50, 25.52, and 25.51 g. The last place can vary by as much as 0.02 g and is only approximately known.

Similarly, in performing any arithmetic calculations, only those figures are retained which are significant. The *least precise measurement* determines the number of figures to be used.

In addition and subtraction, the number of significant figures is determined by the largest absolute uncertainty in any of the measurements.

Examples

Addition

$$\begin{array}{r} 49.6\ \text{g} \\ 3210.\ \text{g} \\ \underline{0.496\ \text{g}} \\ 3260.096\ \text{g} \end{array} = 3260.\text{g}$$

= four significant figures because 3210. is known only to ± one gram

Subtraction

$$\begin{array}{r} 32.9\ \text{g} \\ \underline{0.0496\ \text{g}} \\ 32.8504\ \text{g} \end{array} = 32.9\ \text{g}$$

= three significant figures

$$\begin{array}{r} 83.958\ \text{g} \\ \underline{83.616}\ \text{g} \\ 0.342\ \text{g} \end{array} = 0.342\text{g}$$

= three significant figures

For multiplication and division problems, a different method is used to help evaluate the significance of the last figure. For example:

$$\frac{1.010\ \text{g}}{0.890\ \text{g}} = 1.135$$

In this case the number 1010 is uncertain to 1 part in 1010, and the number 890 varies by 1 part in 890. Therefore, the number of significant figures in the answer will be limited by this. If the answer has only three significant figures, the variation is 1 part in 113, if four significant figures, 1 part in 1134, and if five significant figures, 1 part in 11348. In this case, four figures give a more accurate picture of the limitations of these particular measurements. An example in multiplication is:

$$\begin{array}{rl} 1.6 & \text{cm} \\ \times\, 0.46 & \text{cm} \\ \hline 0.736 & \text{cm}^2 = 0.74\ \text{cm}^2 = \text{two significant figures} \end{array}$$

Zeros may be significant or not. One of the numbers in the examples above is written as 0.890. The zero to the left of the decimal point, which locates the decimal point, is not significant. The zero to the right of the decimal point is significant. It indicates the reliability of the measurement. If a number, like 1000, has no decimal point, then only one significant figure is implied.

To avoid confusion in reporting zeros and their significance, it is often convenient to express numerical values in exponential form. In this way the above number may be reported as 8.90×10^{-1} to indicate three significant figures. Or, for instance, when very large or very small numbers are reported, the exponential form becomes necessary. Avogadro's number is reported as 6.023×10^{23} (four significant figures). You will find almost all your arithmetic manipulations much simpler if you express numbers in exponential form (scientific notation) whenever possible.

Measurements are expressed by a number and its unit. Never separate the numerical value from its dimension. The units are treated exactly as the number and remain with the number through all algebraic operations. For example, to obtain the volume of a cube 2 cm on a side, write: 2 cm x 2 cm x 2 cm = 8 cm^3.

LABORATORY TECHNIQUES

DISCUSSION

A chemist studies pure substances and mixtures: their composition and structure, their reactions, and their physical and chemical properties. So that he can be sure that he does not introduce impurities through carelessness, or endanger himself or those who work with or near him, certain procedures for handling chemicals have been developed. Described below are a few fundamental techniques.

Cleanliness

Before you begin to work in the laboratory, be sure your bench area is clean. While you are working, clean up any spilled chemicals immediately. Wipe your area before you leave and be sure your sink is clean.

All your glassware and equipment should be clean, but they seldom need to be dry. If, however, dry equipment is necessary, there are several acceptable procedures. For other than volumetric equipment, one method is to dry the glassware or porcelain in an oven at 110 to 150°C. Another method is to swirl a small quantity of a water-soluble, volatile liquid, such as acetone, in the vessel to be dried. When the container so treated is inverted and allowed to drain, it dries in a few minutes. These volatile liquids are flammable. Do not heat equipment that has been rinsed with solvent.

Removal of Liquids and Solids From a Vessel

Most liquids are stored in narrow neck bottles with ground glass stoppers. If the stopper is flanged, it can be held between the fingers of the hand that holds the bottle. If the bottle closing is a screw cap or a rubber stopper, it must then be held with the other hand. *Never place the cap or stopper on the laboratory bench.* It may pick up contaminating materials from the desk surface or deposit corrosive or otherwise undesirable chemicals (Figure 1).

Liquids may be poured from a beaker or Erlenmeyer flask without loss by holding a stirring rod against the edge of the container from which the liquid is being poured. Point one end of the rod down into the receiver without touching the walls and carefully pour the liquid down the rod (Figure 2).

Figure 1. Pouring a liquid from a bottle into a beaker.

Figure 2. Liquid transfer using a glass stirring rod.

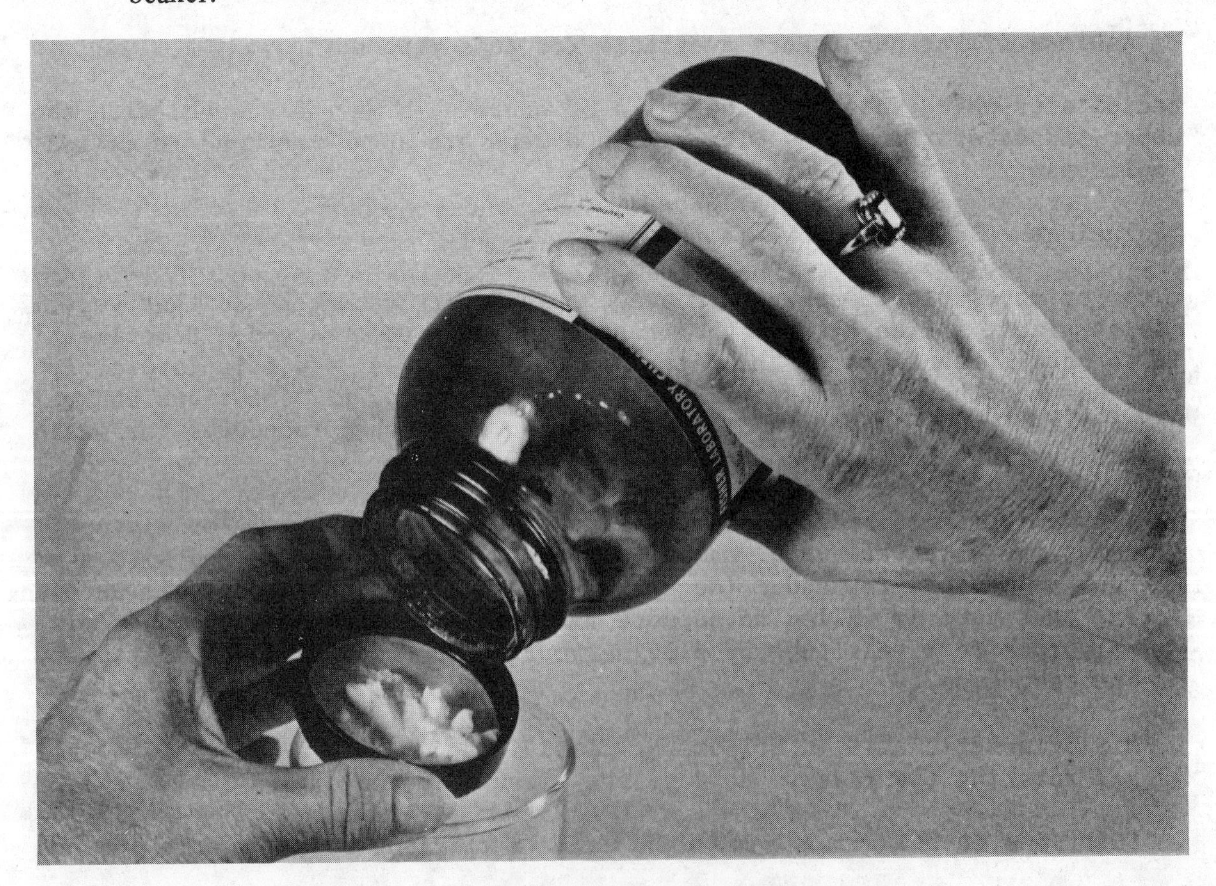

Figure 3. Pouring a solid from a bottle into a beaker.

Decantation

When an insoluble solid is very dense, it settles easily to the bottom of the container. The clear supernatant liquid can be easily poured off with minimum disturbance of the solid. This technique is called decantation.

Solids

Crystalline or powdered reagents are stored in large mouth bottles usually closed with a screw cap. A small quantity of material may be removed from the bottle by tilting and gently rotating it. Again be sure not to put the cap on the bench. The substance may be transferred directly to a container or to the cover of the bottle from which still smaller quantities may be removed (Figure 3).

Never insert a stirring rod or pipet, spatula or scoopula, or any other utensil into a reagent bottle for the removal of chemicals. Such procedures contaminate reagents.

Always replace the stoppers or covers of reagent containers when you have taken what you need.

Never switch the stopper or cover of one reagent bottle with another.

Read the label on the bottle twice, once when you remove the reagent from the bottle and again when you cover the bottle.

Gravity Filtration

Separation of a solid from a liquid may be accomplished by filtering. Filter paper is made in a variety of sizes. The paper, when folded properly (Figure 4) should fit snugly to within 3 mm of the top of the funnel.

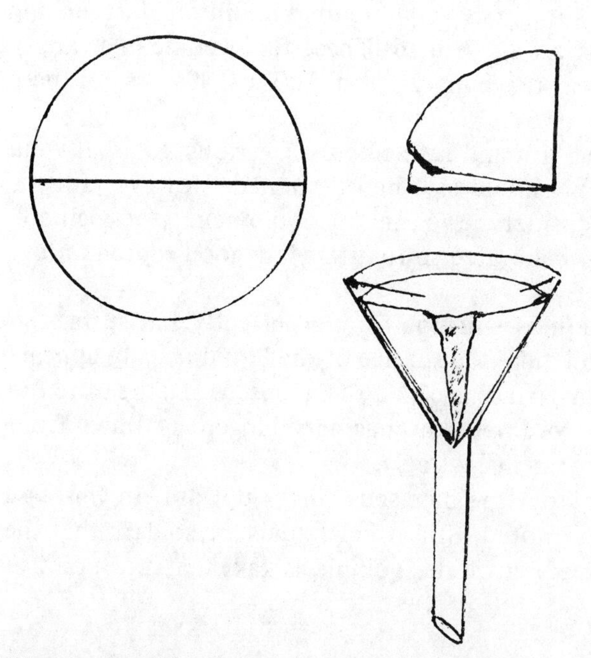

Filter paper folded in half, and then quarters. One corner of fold is torn to make a tit tighter fit.

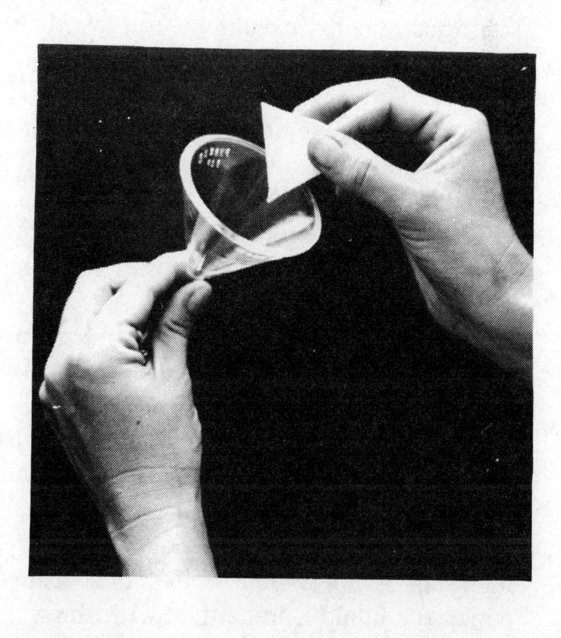

Inserting the properly folded filter paper into a funnel.

Figure 4

Filter paper is also made in a variety of porosities. A fine filter paper, such as Watman No. 42, is used to retain solids of very small particle sizes. Naturally, filtration is slow through such paper. Therefore, when particles of a solid are large, a more porous paper is used, and filtration can be accomplished more quickly.

Precipitates which adhere to the walls of containers may be removed with the aid of a rubber-tipped stirring rod. The tip is a separate attachment and is called a rubber policeman.

Mixing Solutions

In this laboratory manual it must be assumed that, unless other directions are given, all solutions and reacting substances must be thoroughly mixed.

Many small-scale reactions are carried out in test tubes. For mixing solutions in these small, narrow vessels a thin long glass rod with a small knob on the end is used. (Never substitute glass tubing for this purpose.) The test tube is filled to no more than 2/3 of its capacity. Never mix solutions in a test tube by placing your thumb over the mouth and inverting the test tube.

Quantitative Measurement of Liquids

Almost all glassware equipment is stamped with a measure of its capacity. However, such pieces as beakers, Erlenmeyer flasks, Florence flasks, and test tubes are not labeled exactly. For example, a 250-ml Erlenmeyer flask may hold from 275 ml to 290 ml of liquid. Characteristic of volumetric equipment for a quantitative measurement is the constricted neck of the vessel or the long narrow tube. This equipment has finely etched rings, indicating a precise measure. Frequently, the degree of precision of the equipment at a specified temperature will also be stamped on the vessel. Least precise of the volumetric measures is the graduated cylinder. It is less precise because the tube has a relatively large bore. A similar device but with narrower bore and greater precision is the buret. A still more precise volumetric measure is the transfer pipet. The above-mentioned refined equipment is used for measuring specified quantites of solutions ranging in volume from tenths of a milliliter to several hundred milliliters. For making up solutions of definite concentration, a volumetric flask is used. These flasks come in a variety of sizes. Those with which you may work this year range in size from 100 to 1000 ml. All these flasks have a single etched line on the neck.

The above discussion has emphasized the precision and refinement of various kinds of volumetric equipment. The accuracy of this equipment should also be questioned. However, for the purposes of the general chemistry laboratory, we consider the volumetric equipment to be accurate. In other words, we consider that these have been calibrated so that they are good reproductions of the standard.

Because of surface tension and capillarity, liquids will not form a perfectly flat surface in contact with the walls of the container. In a narrow tube the surface of the liquid usually appears to curve downward toward the center. This curved surface is called the meniscus. To measure the volume accurately, the liquid surface must be at eye level. An imaginary tangent is drawn from the lower point of the meniscus to the wall of the vessel (Figure 5).

Some liquids have such high surface tension that the meniscus curves upward. In this case the volume at the highest point of the meniscus is noted. Still other liquids are so dark that the meniscus is not visible. When this occurs, the reading of the volume is taken at the point at which the liquid is in contact with the wall.

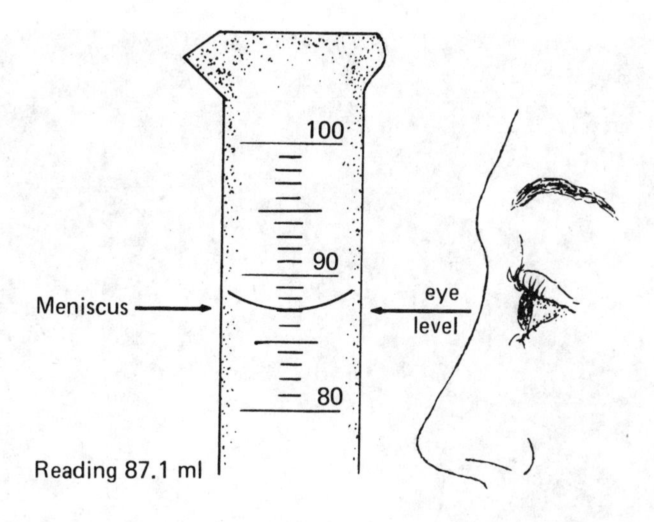

Figure 5. Reading of the meniscus.

It has been previously mentioned that glassware must be very clean. When a liquid is poured from a container, no droplets of moisture should be observed adhering to the glass. Instead, the liquid should drain in a continuous film. This is particularly crucial in the quantitative measure of volumes.

Use of Pipets

Pipets are filled by immersing the tip in the liquid. Reduced pressure at the other end of the pipet allows liquid to be drawn up into it. Since pipets are often used to measure the volumes of liquids that are corrosive and poisonous, *never insert a pipet into your mouth* to draw up the liquid. Your breath can also contaminate both the pipet and the liquid to be transferred. Therefore use the technique described below *at all times.*

Using a trap, connect the long rubber tube attached to the glass T to your pipet (see figure). Put a pinch clamp on this rubber tube. Also put a pinch clamp on the short rubber tube attached to the other leg of the T. Then connect the rubber tube attached to the right-angle bend to the aspirator. Turn on the faucet to reduce the pressure in the system, insert the pipet in the liquid to be transferred, and slowly open the pinch clamp on the long rubber tube connected to the pipet. Use the pinch clamp to control the speed with which the liquid is drawn into the pipet. When the liquid is about 6 cm above the calibrated mark, close the pinch clamp. If you are right-handed, hold the pipet in your right hand, with your index finger close to the top of the pipet in order to be able to put your index finger over the opening at the proper moment. Carefully remove the rubber tube from the pipet with your left hand, and *immediately* close the opening with the index finger of your right hand. The pressure of your finger on the opening controls the flow of liquid. By cautiously adjusting the pressure, allow the liquid to fall slowly to the calibrated mark. Perfection of this technique requires some practice (see Figure 6).

Weighing

Many types of balances may be encountered in the laboratory, for example, the platform balance for weighing large quantities of material (over 600 g); the triple beam balance (Figure 7), for weighing quantities between 0.5 and 600 g; the analytical balance for quantities of less than 100 g that require very precise weight measurement; and the top loading single pan balance (Figure 8) which has been designed in different models depending upon the range desired.

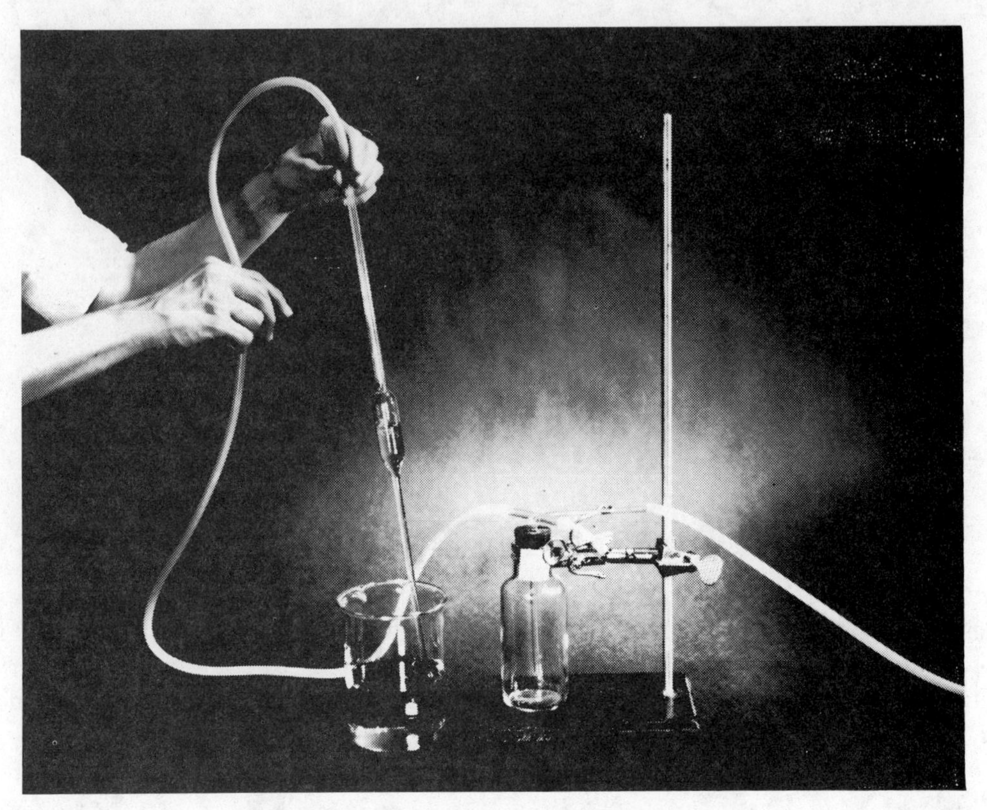

Figure 6. Drawing a solution into a pipet.

The triple beam balance (Figure 7) consists of a single pan suspended on the left side of a beam. The right side of the beam is divided into three sections, each of which holds a rider. The smallest weight (0.00 to 1.00 g) is measured by the beam closest to the operator. The largest weight is held by the middle section. This beam is notched to hold the rider in a given position and is divided into 10-g units ranging from 0.00 to 100 g. The beam furthest from the operator is also notched and measured weights from 0.00 to 10.00 g in 1-g units. Some triple beam balances have two additional weights, each adding 100.0 g to the beam. These are suspended from screws at the right end of the beam. Weighings are made most efficiently by starting with the heaviest weights first and decreasing the load as indicated by the deflection of a pointer at the end of the beam along a marked scale. The scale is marked off in units so that it is possible to determine when the weights are balanced by the equal deflection of the pointer on either side of the zero mark. To obtain the weight of the object, all the weights on the beam are totaled.

Comparable in precision and use to the triple beam balance is the top-loading single-pan balance (Figure 8). This balance has two arms of unequal length. The shorter arm supports the weighing pan and all the weights for which the balance was designed. The longer arm is an exact counterbalance. When an object is weighed, it is placed on the weighing pan attached to the short arm. The weights are removed until the two arms are again exactly counterbalanced. This means that the weights are subtracted from the side supporting the object being weighed, rather than being added to the other side as in most manual balances. The weight of the object is read out directly on optical scales on the front of the machine; in some cases by a direct digital readout; in others, fine readings may be made by other devices, such as a vernier scale. These balances are exceptionally rugged and efficient.

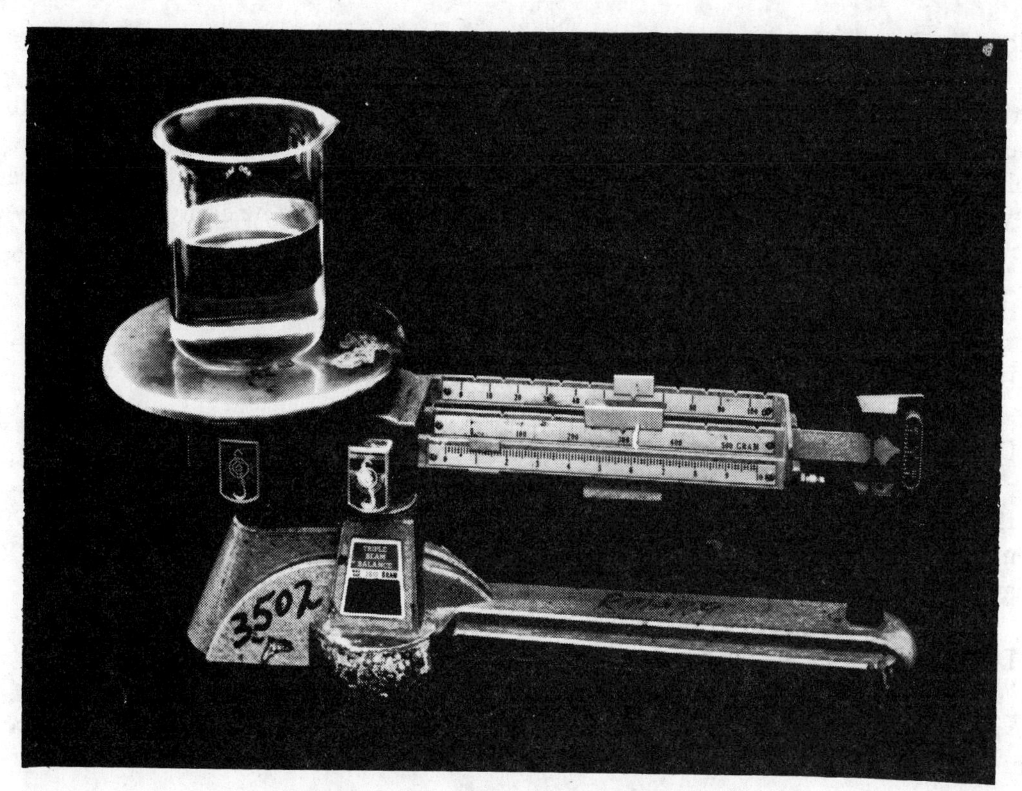

Figure 7. Triple beam balance.

Figure 8. Single pan balance.

Precautions To Take in Weighing

1. A triple-beam or platform balance contains two agate knife edges, one upon which the pan is suspended and one upon which the beam rests. The single-pan automatic balance has two sapphire knife edges. These knife edges are precisely ground so that the balance can swing freely with a minimum of resistance caused by friction; they can be easily damaged by careless and rough handling. Therefore:

   a. Do not remove the balance from its location.

   b. Do not place any warm or hot object on the balance pan.

   c. Do not drop objects on the pan.

   d. If the balance beam has an "arrest", use it to stop the motion of the beam before removing the object from the pan. This is of *utmost* importance in the operation of the analytical balance.

   e. Do not remove objects from the pan with sudden motion.

2. The pan of the balance is removable, but it is calibrated for that particular balance. Therefore:

   a. Do not remove the pan from the balance.

   b. Do not exchange the pan of one balance for that of another balance which may appear similar.

3. Reagents can corrode the balance pans. Therefore:

   a. *Never* weigh any reagent directly on the pan.

   b. *Always* use a glass or porcelain container.

4. Should there be any apparent difficulty in the operation of the balance, an instructor should be consulted. *Do not attempt any adjustments.*

5. A slight difference in the accuracy of several balances is usual. This does not offer any difficulty, however, if the same balance is used for all the weighings in a given experiment.

Heating Solutions and Reacting Mixtures

All Pyrex and porcelain equipment *except* the graduated glassware can be heated over a Bunsen burner flame. Beakers and flasks are placed on an asbestos wire gauze to ensure even heat distribution across the bottom of the container. In addition, they are not filled over two-thirds of their volume, and either boiling chips or a glass stirring rod are added to the liquid to prevent "bumping" (sudden boiling of a solution because of superheating). A test tube is usually held at an angle in a test-tube holder directly over the flame. It is moved slowly back and forth through the flame to prevent the solution from being overheated in a particular spot. Because of the narrowness of the test tube, bumping can cause the ejection of the entire contents from the tube.

For this reason it is wise *not to look down into a test tube that is being heated and not to point the open end at those working near you.* Solutions in test tubes are often more conveniently and evenly heated in water baths.

Glassware which should not be heated over a flame includes bottles and soft glass, as well as volumetric glassware.

The Bunsen Burner

The Bunsen burner is most commonly used as a source of heat in the laboratory; it is a good idea to become well acquainted with its characteristics.

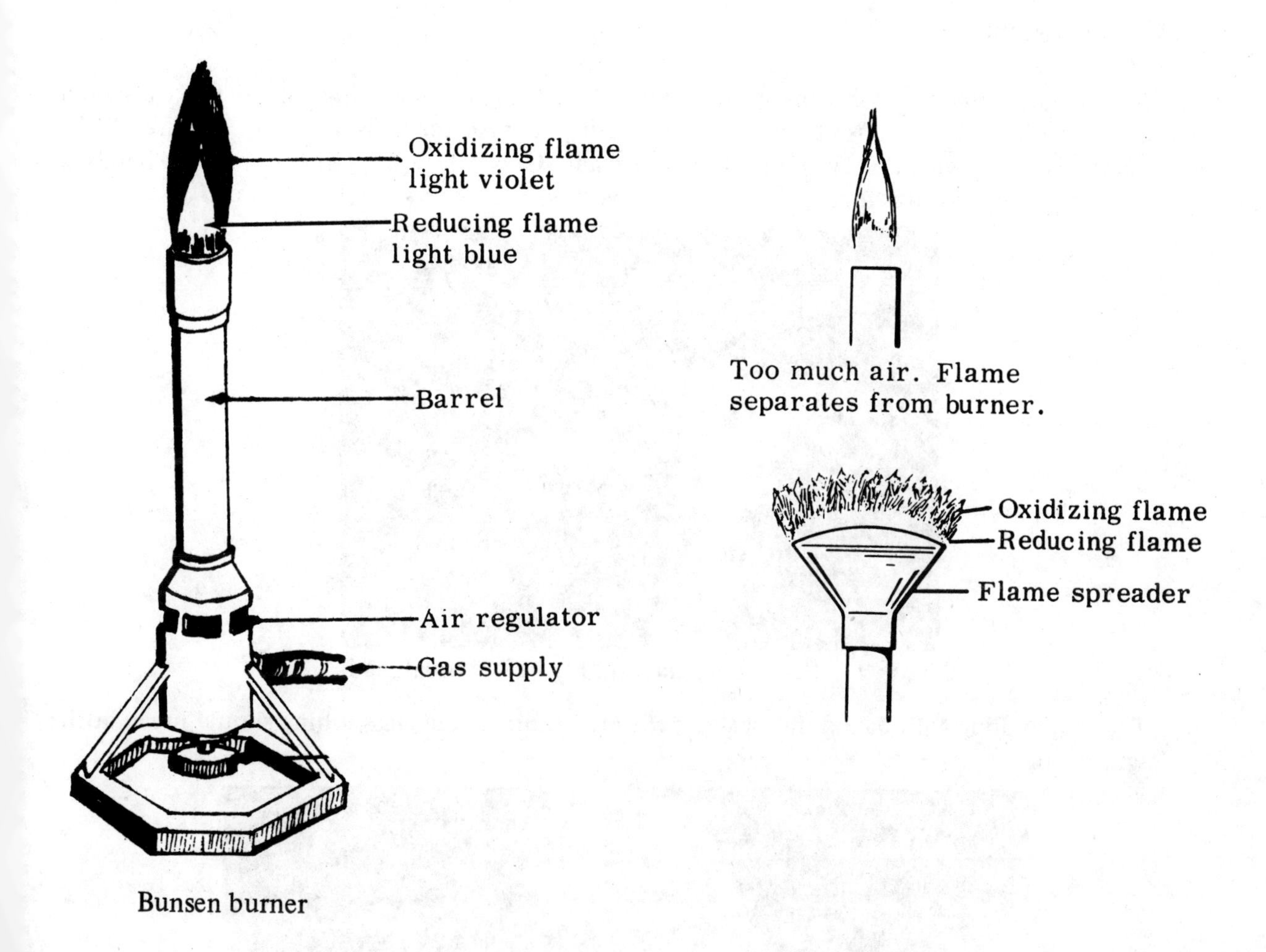

Bunsen burner

The burner consists of a barrel at the base of which are air vents which may be opened or closed. Gas is introduced to the burner below the air vents through a rubber hose connected to the outlet on the desk. By regulating the flow of gas at the outlet, and by opening and closing the air vents, the gas can be made to burn efficiently. The temperature of parts of the flame varies. The hottest part of the flame is the point of the light blue cone.

To obtain a visual demonstration of this, insert a copper wire into the base of the flame and slowly raise the wire to the top of the flame. Observe the physical changes in the wire.

Glass Manipulation

Caution

Glass has the same appearance whether it is hot or cold. To save yourself the misery of finger burns, place all glass with which you have been working in the wire rack or on wire gauze until it is cool. This will also prevent the hot glass from cracking, because it will not be in contact with the cold laboratory bench top. It is also advisable not to have windows open near glass working operations. The draft is sufficient to cool both glass and flame to the point at which the glass is too hard to be workable. Do not put hot glass into wastebaskets, because paper or chemicals may become ignited.

1. To cut glass tubing:

   Lay the glass tube to be cut on the laboratory bench. With your triangular file make a scratch on the surface of the glass at the point at which you wish the glass to break (Figure 9). The scratch is made with one firm stroke. *Do not saw.* Grasp the tubing firmly in both hands so that your thumbs are together on the opposite side of the scratch (Figure 10).

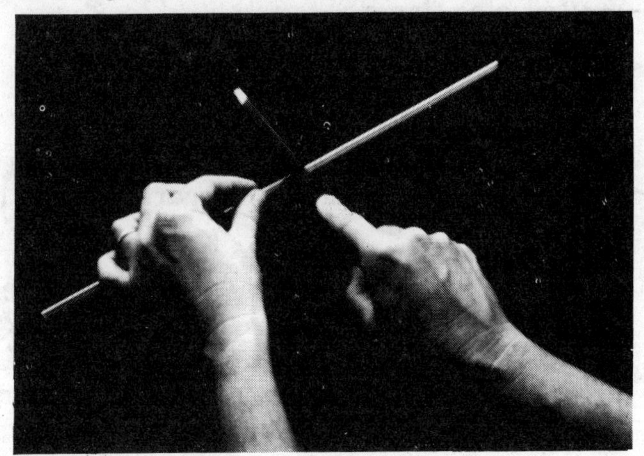

Figure 9. Scratching glass tubing before breaking it.

   Then, by pulling with hands and pushing with the thumbs, the glass tubing should break at the scratch line.

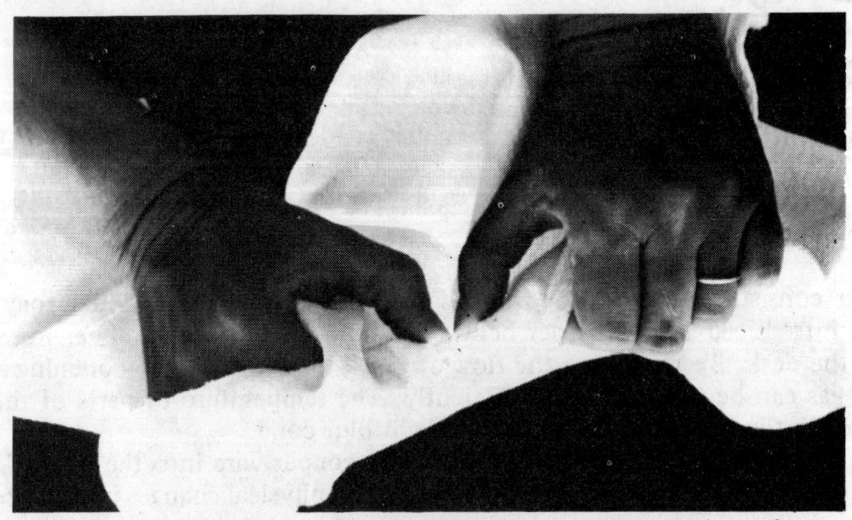

Figure 10. Preparing to break the tubing at the scratch mark.

2. To fire-polish tubing:

   Adjust your Bunsen burner flame so that you have a bright blue cone of flame inside a light violet cone. Insert the end of the tube to be fire-polished in the flame so that the tip is just inside the outer violet cone. Rotate the glass so that it will heat evenly. In a few seconds the glass will have become hot enough to flow slightly, rounding the ends of the glass. Do not heat so long that the tube collapses.

3. To smooth jagged edges of tubing:

   Occasionally, tubing does not break evenly. To correct this, hold the tubing firmly in one hand down over your wastebasket. Holding your wire gauze in the other hand, and using short sharp strokes, chip off the jagged edges of the glass. Then fire-polish the edges.

4. To draw a tip for a dropper:

   Using both hands to hold a section of glass tubing, heat and rotate the glass continuously in the flame, making sure that both left and right hands rotate at the same rate of speed. Since the flame is hottest at the point of the light blue cone, be sure to keep the glass at this point. When the glass has become very soft and pliable, *remove the tubing* from the flame and *then pull slowly* until the tube has the desired thickness; immediately hold the tube vertically to allow gravity to help you obtain straight tips. When cool, the constriction is cut where it has the desired thickness. An effective technique for cutting thin-walled glass tubing is to place a triangular file on your working surface. Then lightly, and with one gentle sweeping motion, slide the tubing across the edge of the file. After a few practice cuts, you will know exactly how much pressure to use to produce a visible scratch.

5. To bend glass tubing or rod:

   To bend glass tubing or rod successfully without folding it, use your flame spreader. This elongates the flame horizontally and provides a long flat light blue area. Adjust the flame so that you have again a light blue area inside a violet flame. Hold the glass firmly in both hands so that it is just above the light blue area. Rotate the glass slowly to ensure even heating. When the glass is *very* soft, remove the glass from the flame and bend it slowly so that the bend will not be distorted by the downward pull of gravity. It may be bent to any desired angle.

6. To insert fire-polished glass tubing into a hole in a rubber stopper:

   Hold the stopper firmly in one hand. With the other hand, wrapped in a towel, hold the glass securely, close to the fire-polished end which is to enter the stopper. By gently pushing and twisting, the glass can be inserted into the stopper. It may be helpful to lubricate both the stopper and glass with water or glycerine.

7. To make the stirring rod:

   Obtain glass rod cut to desired length from the stockroom. Using your Bunsen burner without the wing top, heat the end of the rod until it becomes soft and appears to drip. The end may be shaped by pressing the hot end on an asbestos board.

NAME______________________ SECTION__________ DATE__________

Experiment 1

DETERMINATION OF THE WATER OF HYDRATION

PRELIMINARY QUESTIONS

1. Distinguish between precision and accuracy.*

2. Does good precision always mean a high degree of accuracy? Explain.

3. To what precision is one able to weigh
   a) on an analytical balance? ________
   b) on a top loading centigram balance? ________

4. Explain the purpose of significant figures (digits) in reporting measurements.

5. What is wrong with the following statements?
   a) a sports reporter *estimates* the crowd in a football stadium to be 25,466 people.

   b) the bank reports your balance to be *about* $2,000.

* Discussion of the ideas concerned in these preliminary questions can be found in the introductory pages on *Experimental Limitations*.

6. Assume that each of the following numbers is related to a physical measurement. For each value indicate the number of significant figures by expressing it in scientific notation. How many significant figures are in each value?

   a) 7000 kg_____ c) 0.00700 km_____ e) 703.0 g_____

   b) 200.30 cm_____ d) 703 g_____

7. Express the results of the following arithmetic operations with the proper number of significant figures. Assume that each value is uncertain in the last digit.

   a) 3.43 g + 100.2 g + 33 g __________

   b) 77.9 cm x 22 cm __________

   c) $\frac{7.85 \times 10^{-2} \text{ g}}{5.3 \times 10^{-3} \text{ mole}}$ __________

8. Two students are asked to measure the dimensions of the same cylindrical bottle to see if it will hold 75 cm^3 of water. Each is able to measure the dimensions of the bottle to the nearest 0.1 cm. Despite this, the first student reports that the bottle is 3 cm in diameter and 10 cm tall. However, the second student reports that the same bottle is 3.2 cm in diameter and 10.5 cm tall. To calculate the volume of the bottle use $\pi r^2 h$.

   a) What volume will the first student calculate for this bottle?_______

   b) What volume will the second student calculate for this bottle?_______

   c) Will this bottle hold the 75 cm^3 of water?_________

9. Your calculator gives you a number 1761.2868 as the volume (in cm^3) of a box 2 ft by 6.5 cm by 1.75 in. Since this number should reflect the precision of the physical measurements, how many significant figures will you record for the volume of the box?

Experiment 1

DETERMINATION OF THE WATER OF HYDRATION

IDEAS

Gravimetric analysis is based on the weight changes of the system under investigation. The determination of the percentage of water in a hydrate is a type of gravimetric analysis.

A carefully weighed amount of unknown hydrate is heated to drive off the water of hydration. After several heatings, when constant weight has been attained, the difference between the original weight and the final weight represents the mass of water which has been eliminated. It is, of course, assumed that only water has been volatilized and no decomposition of the compound has taken place.

Many inorganic substances contain a fixed amount of water of hydration, or water of crystallization, as for example $CuSO_4 \cdot 5H_2O$. This water is an intimate part of the crystal structure and removing it can alter the spatial arrangement of the ions of the salt. Barium chloride dihydrate crystallizes in the monoclinic form while anhydrous barium chloride is orthorhombic.

Removing the water of crystallization can also cause color changes in the salt. The hydrated copper sulfate is blue and the anhydrous form is white. Even the number of hydrated water molecules can alter the color of an inorganic compound. $NiSO_4 \cdot 7H_2O$ is green and $NiSO_4 \cdot 6H_2O$ is blue-green.

Water can also be trapped in solids in varying amounts depending on the method of preparation. It can be absorbed on the surface of the solid or trapped inside the crystal. In this experiment it is assumed that the unknown sample contains only water of crystallization.

To remove molecules of water from a crystal requires energy to break the bonds which hold the water in the crystal. The strength of these bonds varies enormously. Some solids lose their water spontaneously in a dry atmosphere at room temperature; others require prolonged heating at temperatures well above 100°C. Care must be exercised on prolonged heating that no decompsition other than the removal of water is taking place.

This experiment is also designed to introduce you to the proper use of significant figures and to acquaint you with the appropriate conventions in presenting scientific data. Your report should be complete enough so that anyone reading your results can understand the procedure for the data collection and can reproduce the calculations involved. Measured data must carry the proper units. You should also understand the sources of error in this analysis which will affect the accuracy and precision of your results.

It is important to weigh your samples when they have cooled to room temperature. Weighing hot or warm objects causes error in the recorded weight. Due to changes in air density, convection currents affect the buoyancy of the sample. In addition, unequal expansion of some of the metal parts of the balance may occur.

INVESTIGATION

Purpose: To determine the percent water in a hydrate; to observe the changes in physical appearance of the crystal as water is removed; to learn the operation of the balance; and the use of significant figures in recording data and in reporting the resulting calculations.

Equipment:

- two crucibles and lids
- clay triangle
- top-loading balance
- asbestos squares
- tongs
- bunsen burner
- tripod or ring and ring stand

Chemicals: unknown hydrate

Procedure:

Always use the same balance for consecutive weighings. If there is any error because of a mechanical deficiency of the balance, each weighing is affected by it. However, when the difference between two weighings is taken, this systematic error is cancelled.

Your instructor will tell you whether you are to do two trials consecutively or simultaneously. If you are doing two trials simultaneously, then you must distinguish between the crucibles being used for each trial.

Heat a clean crucible and lid to redness for a few minutes. Cool on an asbestos square to room temperature and weigh on the top loading balance. NOTE: Do not handle the crucible and lid with your hands. Grasp the crucible *on the outside* with tongs.

Question: To what precision can you determine the weight of the crucible?

Record the weight of the crucible and cover on the data sheet. Heat the crucible and lid again for about one minute; cool, weigh and if this weight agrees with the first weighing to within a range of 0.02 gm then constant weight* has been attained. You may be able to report to more significant figures depending on the precision of the balance which you are using.

Put about 1.5 to 2 grams of the unknown hydrate into the crucible and weigh, with cover on, precisely. Replace the crucible on the clay triangle; partially cover with the lid to leave a small opening for the steam to escape.

Heat gently at first to avoid spattering the crystals. When it appears that no more water is escaping, heat more strongly for a few minutes. The entire heating procedure may take 15 to 20 minutes.

Remove the crucible to the asbestos square with the tongs, cover completely with the lid and allow to cool to room temperature. Weigh the crucible with the lid on and record this weight.

*At least two weighings should agree to within the precision of the balance in order to be confident that the measured weight is as close as possible to the true weight and that all volatile impurities have been eliminated.

Return the crucible to the tripod for a second heating of about five minutes. Make certain that during each heating process the lid of the crucible is slightly ajar. Cool,** weigh again and record this weight. This weighing should agree with the first to within 0.02 grams. If it does not, then repeat the heating, cooling and weighing until constant weight is attained.

Repeat this procedure with a second sample of your unknown hydrate.

Calculate the percentage of water in the hydrate for your two trials and determine the average.

**If available, allow the crucible and cover to cool in a desiccator. If this procedure is followed, do not cover the crucible. Rest the lid alongside the crucible inside the desiccator.

NAME________________________SECTION____________DATE____________

DETERMINATION OF THE WATER OF HYDRATION

DATA

	TRIAL 1	TRIAL 2
Weight of crucible + lid + hydrate	______	______
Weight of crucible + lid	______	______
Weight of hydrate	______	______
Weight of crucible + lid + anhydrous compound (first heating)	______	______
Weight of crucible + lid + anhydrous compound (second heating)	______	______
Weight of crucible + lid + anhydrous compound (final heating, constant weight)	______	______
Weight of water lost (a-f)	______	______
Percentage of water in the hydrate	______	______
Average percentage (show calculations)	______	

THOUGHT

1. Why is it necessary to weigh an anhydrous substance with a cover on?

2. What is the specific reason for weighing to constant weight in this experiment?

3. How many significant figures may you use to express the percentage of water in the hydrate?

4. Why is it necessary to allow the crucible and contents to cool to room temperature before weighing?

5. What effect would the following errors have on the accuracy of an experimental determination for the percentage of water in a hydrate? (i.e. higher, lower, or the same as the accepted value).

   a. Some of unknown is spilled before it is weighed in the crucible.

   b. Some solid is lost during the heating.

   c. The anhydrous compound absorbs moisture from the air while it is cooling.

NAME____________________ SECTION__________ DATE__________

Experiment 2

DETERMINATION OF AVOGADRO'S NUMBER

PRELIMINARY QUESTIONS

1. What is the definition of Avogadro's number?

2. Suppose that you have a micropipet which delivers 100 drops per cubic centimeter, and a solution of 5.0% by volume of oil in alcohol. If you had used 10 drops of this solution, how many cubic centimeters of *pure oil* would be left after evaporation of the alcohol?

3. Why must you calibrate your micropipet with oleic acid solution and not with water?

4. How would you calculate the thickness of a liquid layer if you know the volume and the area of this layer?

5. Given a solution of 5.0% by volume oleic acid in alcohol: Two drops (100 drops/cm^3) are placed on a water surface and a continuous, circular film is formed 20 cm in diameter. After all the alcohol has evaporated and the oil is distributed evenly on the surface, what would be the thickness of the layer of pure oil?

6. Suppose you have a box 1 in. by 2 in. by 2.5 in. filled with lump sugar. If you dumped the lumps of sugar from the box and arranged them in one rectangular layer, the area would be 51.3 cm^2. What would be the thickness of one lump of sugar? If you assume that the lumps of sugar are cubic, approximately how many lumps of sugar were in the box?

Optional. Here is a challenging problem.

7. The giant tanker Torrey Canyon was wrecked off the southeast coast of England in March 1967. About 80,000 tons of the 120,000 tons of crude oil which it carried poured into the sea, spreading out to pollute the surrounding coast. Not only the nearby coast was affected, but some of the oil was observed near the French shore.

   The 80,000 tons of oil represented 24 million gallons. There are 3.8 liters per gallon. Calculate the area that would be covered by the oil if the following assumptions are made: A monomolecular layer is formed; each molecule is a cube; the average molecular weight of crude oil is about 400u. If all the oceans and seas in the world cover an approximate area of 140 million square miles, what fraction of the oceans would be covered by this film of oil?

Experiment 2

...TERMINATION OF AVOGADRO'S NUMBER

IDEAS

Avogadro was honored for his contributions to science by the use of his name with a number that has been accepted as a standard in chemistry. In a paper published in 1811, Avogadro hypothesized that equal volumes of gases under conditions of equal temperature and pressure contain equal numbers of molecules. Gay-Lussac had found that gases reacting chemically under conditions of equal temperature and pressure combine in a ratio of small whole numbers. Although many were puzzled by the apparently strange results of Gay-Lussac's experiments, Avogadro's solution to this problem was rejected by the majority at that time. Neglected for about fifty years, Avogadro's hypothesis was revived by Cannizzaro. It was accepted then because evolving ideas on the nature of gases yielded powerful evidence in its favor. This phase of the development of ideas in chemistry has been termed "the final stage of the chemical revolution of the eighteenth and nineteenth centuries; the result was the development and sharpening of many of the fundamental concepts upon which modern chemistry is built."*

The acceptance of Avogadro's hypothesis enabled the determination of the relative weights of many atoms and molecules; atomic weights and molecular weights were then defined in terms of the accepted standard, 0 = 16.0000 amu. The gram atomic weight and the gram molecular weight were defined as the standard "package", or mole, of atoms or molecules. The name "Avogadro's number" was given to the number of molecules or atoms in a mole. This number, 6.02×10^{23}, is now referred to as a "mole" in the same manner that the number 144 refers to a gross.

Avogadro's number has been determined by various methods, all of which yield the same results, within the limits of experimental error. The method we shall use today, although primitive, yields surprisingly good results which are of the right order of magnitude if the experiment is carried out carefully. It utilizes an interesting property of certain large molecules such as fatty acids. If a drop of fatty acid is placed on the surface of water, it will spread out to form an extremely thin film. Observations of this sort were recorded as long ago as 1773 by Benjamin Franklin, who noted that 1 teaspoon of oil spread out to form a film of about 22,000 square feet on a pond near London. Franklin was investigating an old observation—the effect of pouring oil on troubled waters.

That this extremely thin film is probably the thickness of one long-chain molecule may be demonstrated by placing a wire across the surface of a shallow container filled to the brim with water, and allowing a drop of oil to fall on the water to one side of the wire. The oil will spread out over the water surface and attach itself to the wire and to the edges of the container because

* G. Holton and D. Roller, *Foundations of Modern Physical Science*, Addison-Wesley Publishing Company, Reading, Mass., 1958, p. 400.

of intermolecular forces. If the wire is moved to stretch the film, the film breaks in places, and islands of water are visible. Obviously, the film cannot be stretched indefinitely; it breaks up as the wire continues to pull it after it has achieved minimum thickness of the thickness of one molecule.

In this experiment, the fatty acids used must be quite dilute. One drop of pure oleic acid will cover a water surface of about 200 m^2.

The reasons for this interesting behavior of large insoluble molecules have been attributed to the forces existing between the polar water molecules and the functional groups of the large molecules, thus causing a vertical or near-vertical orientation of the long molecules in relation to the water surface. Intermolecular forces between the carbon atoms account for the horizontal cohesiveness of the film.

Stearic or oleic acid, because of their strong polar group, large intermolecular forces of cohesion between the carbons, and their uncomplicated straight chain structure, are often used to study monomolecular films. The thickness of the film, which is the thickness of one molecule, may be calculated from a knowledge of the volume of one drop, the mass of oleic acid in one drop, the density of oleic acid, the volume of oleic acid comprising the film, and the area of the film. If the simplifying assumption** is made that the molecules are cubes, then the volume of one molecule may be calculated. From the known density of oleic acid, the mass of the molecule is calculated; and finally from the gram molecular weight, Avogadro's number is determined.***

INVESTIGATION

Purpose: To determine the number of molecules in a mole by casting a thin film of oil and assuming the formation of a monomolecular layer.

Equipment:
- 1 stainless steel or white enamel pan or cafeteria tray (if an enamel pan is used, its surface must be unbroken)
- 1 medicine dropper bulb
- 1 micropipet
- 1 10 ml Erlenmeyer flask and stopper

Chemicals:
- lycopodium powder
- 0.5% (by volume) solution of oleic acid in methyl or ethyl alcohol

Procedure:

In this experiment the importance of immaculately clean equipment cannot be overemphasized. A small amount of grease from your finger will result in unreliable values.

Draw glass tubing into a capillary so that 100 to 150 drops/cm^3 of the 0.5% solution of oleic acid are delivered. Withdraw about 5 ml of the oleic acid solution from the stock bottle and place in a small clean dry Erlenmeyer flask. Keep this closed with a stopper. Determine the volume of 1 drop of 0.5% solution of your sample of oleic acid.

** Recent studies using the film balance and the electron microscope have shown that the molecules actually exhibit all orientations on the water surface, and that they are not uniformly packed unless compressed.

*** References for additional discussion: L. Carrol King and E. K. Neilson, *J. Chem. Educ., 35,* 198 (1958).

Add deionized water to the pan or tray to a depth of several centimeters. Evenly dust the surface with a very thin layer of lycopodium powder. The lycopodium powder makes the boundaries of the oil film easily visible. Discard the first drop. Then put 1 drop of solution on the surface of the water and measure the diameter of the film in two directions at right angles. Average and record these values. The alcohol evaporates, leaving a layer of oleic acid.

NAME______________________________ SECTION__________ DATE________

DETERMINATION OF AVOGADRO'S NUMBER

DATA

Show all methods of calculation clearly.

a. Number of drops/cm^3 of 0.5% oleic acid solution delivered ________

b. Volume of one drop ________

c. Volume of oleic acid in one drop ________

d. Diameter of film ________

e. Area of film ________

f. Thickness of film, $\frac{\text{(vol)}}{\text{(area)}}$ ________

g. Volume of one molecule (assuming molecules are cubes) ________

h. Density of oleic acid 0.89 g/cm^3

i. Molecular weight of oleic acid ($C_{18}H_{34}O_2$) ________

j Molar volume ________

k. Number of molecules in one mole (assuming molecules are cubes) ________

The oleic acid molecules are actually 10 times longer than their width. So, to correct for the error made by assuming the molecule to be a cube, another assumption must be made--that the molecules stand vertically and are in contact. As noted previously, this assumption has not been found to conform to the experimental evidence. Nevertheless, the use of this simplifying assumption seems to improve the calculated value for Avogadro's number. In that case, since the volume of the molecule has been calculated from the thickness of the layer or the length of the molecule, the value for the volume will be too large, by a factor of 100. Avogadro's number will therefore be too small, also by a factor of 100.

l. Number of molecules in one mole (assuming molecules are rectangular) ________

m. Accepted value of Avogadro's number ________

THOUGHT

1. What are the assumptions made in this method of calculating Avogadro's number?

2. What are the possible experimental sources of error in carrying out this experiment?

3. Which of the above experimental measurements is most subject to uncertainty?

NAME______________________________________ SECTION____________ DATE___________

Experiment 3

BALMER SERIES FOR HYDROGEN

PRELIMINARY QUESTIONS

1. a) What theoretical model accounts for the emission spectra of atoms?

   b) How do emission spectra of atoms originate theoretically?

2. a) Define nanometer
   b) How many nanometers are in one centimeter?

3. What is the mathematical relationship between frequency, wavelength and speed of light?

4. a) Give an equation for the relationship between the spacing of the energy levels of electrons in an atom and the wavelength of the observed lines in an emission spectrum.

   b) If light of 615.2 nm is emitted, to what frequency of light does this correspond.

   c) To what energy level difference would this wavelength correspond?

Experiment 3

BALMER SERIES FOR HYDROGEN

IDEAS

Atomic line spectra for many elements were observed more than 100 years before an adequate theory was formulated to account for their origin. It had been observed that light is emitted by gases or vapors in a tube through which an electric spark or current passes, or by volatile solids sprinkled into a non-luminous flame. For instance, a stable salt sprinkled into the flame produces a yellow color; neon glows reddish in an electric discharge tube.* Permitting the light so produced to pass through a prism or grating spectroscope revealed the presence of a series of bright lines separated by dark spaces. Unlike gases, light produced by heating a solid to glowing is the same for all solids, the color depending only upon the temperature; passing that light through a spectroscope produced a continuous spectrum of colors, one blending into the other.

Pioneering investigations of the spectra of gases had been done as early as 1752 by the Scottish physicist Thomas Melvill who wrote, "having placed a pasteboard with a circular hole in it between my eye and the flame . . . I examined the constitution of these different lights with a prism." Each element studied in this way revealed its own unique set of bright lines of specific wavelengths which was later called the line emission of atomic line spectrum. This new method of identifying the elements, called spectrum analysis, spurred the production of finer instruments. The physicist G. R. Kirchhoff and the chemist R. W. Bunsen observed new emission lines in mineral water vapor, leading to the discovery of two new elements. cesium (1860) and rubidium (1861). Studies of emission spectra of meteorite vapor revealed that they contained only elements found on earth. The heavens and earth were so united, being composed of the same building materials.

In addition to continuous and line emission spectra, a third kind called the absorption line spectrum had been discovered in 1814 by Joseph Fraunhofer. He found that the spectrum of the sun and some bright stars appeared to be crossed by a series of dark vertical lines; he counted more than 700 in the sun's spectrum, and used the letters of the alphabet to identify the most prominent lines. One clue to the interpretation of the dark lines was the observation that sodium vapor produced two bright yellow lines that corresponded exactly in wavelength and position to two prominent dark lines (called D lines) in the solar spectrum. By 1859, Kirchhoff had performed many laboratory investigations on dark line spectra. For instance, he allowed light from a glowing solid to pass through a glowing vapor, e.g., a flame sprinkled with sodium chloride, and he found the absorption lines that corresponded exactly in wavelength and position to the emission lines, and just as they appeared in the sun's absorption line spectrum. These experiments led to the

* A discharge tube is a glass tube containing a desired gas at low pressure through which a very high voltage can be applied.

conclusion that the dark lines in the sun's spectrum represent the apparent absorption of some wavelengths by the cooler glowing gases surrounding the hot dense core of the sun. The core, composed of gases under great pressure, produces a continuous spectrum like that of a glowing solid. In passing through the gaseous envelope, Kirchhoff wrote that the dark lines "occur because of the presence of these elements in the glowing atmosphere of the sun which would produce in the spectrum of a flame bright lines in the same position. We may assume that the bright lines corresponding with the D lines in the spectrum of a flame always arise from the presence of sodium; the dark D lines in the solar spectrum permit us to conclude that sodium is present in the sun's atmosphere." Here was additional evidence that the heavens and earth are composed of the same elements. This interpretation was also a key to the chemical make-up of the sun and other stars, and an additional tool for chemical analysis on the earth.

The great variety of line emission spectra and the apparent lack of relationship among the spectra of elements, even those close to each other on the Periodic Table, was puzzling. For instance, as mentioned earlier, sodium vapor showed only two bright lines while iron vapor showed more than 6000.

Early attempts to produce order from the accumulated wealth of spectral data were made by many scientists, including J. J. Balmer, a Swiss school teacher who carefully studied the hydrogen line emission spectrum. In 1885, by trial and error, Balmer formulated an empirical equation which related the wavelengths of the four lines then visible. In modern notation, the formula is

$$\frac{1}{\lambda} = R\left(\frac{1}{2^2} - \frac{1}{n^2}\right)$$

with n having the values of 3, 4, 5, or 6 and R, the Rydberg constant, a value of 109,678 cm^{-1}. This accounted for the four visible lines.

Although the equation at the time it was formulated had no theoretical meaning, Balmer imaginatively went on to speculate that perhaps the 2 might also be replaced by other integers. The general equation would then be

$$\frac{1}{\lambda} = 109{,}678\ cm^{-1} \left(\frac{1}{n_i^2} - \frac{1}{n_o^2}\right)^*$$

where n_i could be 1, 2, 3, 4, or 5 and n_o would be $n_i + 1$, $n_i + 2$, and so on. These series were eventually found, three in the infra-red ($n_i = 3, 4,$ or 5) and one in the ultra-violet ($n_i = 1$).

The hydrogen spectrum was an important key in the attempts by Niels Bohr to find a solution to the mystery of atomic structure.

In 1911, Niels Bohr, having just acquired his Ph.D. in his native Denmark, arrived in England for post-doctoral work in Sir J. J. Thomson's Cavendish Laboratories at Cambridge University. However, Bohr's insistence on applying the newly formulated quantum theory to the problem of the structure of the atom clashed with Thomson's more classical approach. After a short time, Bohr left the inhospitable environment of Cambridge and was fortunate in securing a post with Ernest Rutherford at the University of Manchester. Here, exciting work was going on involving the scattering of alpha particles by thin metal foils. The results of these studies pointed to a nuclear atom, quite at variance with Thomson's model. Thus, the atmosphere at Manchester was more congenial for Bohr and his ideas.

It was here that Bohr's intuition that there must be a connection between Einstein's and Planck's discrete light quanta and the discrete line spectra of atoms matured into the quantum theory of the atom.

* The subscript "i" means inner energy level; "o" means outer energy level.

Planck had suggested the revolutionary idea that energy is emitted and absorbed in the submicroscopic world of atoms in discrete packets called quanta. His famous equation, $E = h\nu$ (or hc/λ) giving the energy value of these quanta was utilized by Einstein to create the concept of packets or quanta of light, to which the name photon was given. Bohr made use of these contributions and other concepts to derive the Balmer equation by theoretical means and to account for the bright line spectrum of hydrogen.

Each bright line, according to Bohr's model, represents large numbers of photons emitted when electrons drop from higher to lower (outer to inner) permitted energy levels. The allowed energy levels are termed quantized. The n values which appear in the equation for the hydrogen spectral series signify the permitted quantum levels that the electron may occupy. These n values can vary from $n = 1$ to $n = \infty$ only by whole numbers. For $n = 1$, called the ground state, the energy level is the lowest; the electron is closest to the nucleus.

Even though the Bohr formulation holds only for hydrogen (a one-electron system), his pioneering work was one of the main foundations on which the successful quantum mechanical treatment of the atom is based. His important contribution was universally acknowledged when he received the Nobel Prize.

INVESTIGATION

Purpose: To determine the wavelengths of 3 lines of the hydrogen spectrum; to calculate the Rydberg constant from the measured wavelengths; to calculate the energy level differences, ΔE, from the measured wavelengths.

Equipment:

- visual spectrometer
- Tungsten lamp
- Mercury arc
- Hydrogen lamp

Procedure:

I. Examination of the Visual Spectrometer

a. In Figure 11, the source slit 'A' is at the end of the right tube. The screw 'C' permits variation of the slit width. With this the thickness and resolution of lines can be varied.

b. The spectral light source 'B' is a discharge tube which contains a gaseous sample of the element being studied. This gas glows when voltage is applied. The tube is placed in front of the slit.

c. The upper tube contains at its end a frosted glass permitting light from a tungsten lamp to enter and light up the scale (D) which is to be read to the *tenths* place. This scale must be calibrated with known lines from the mercury spectrum* before it can be used to determine the wavelengths in the hydrogen spectrum.

---

* We are assuming that the wavelengths of the mercury spectrum have been carefully determined and are known accurately.

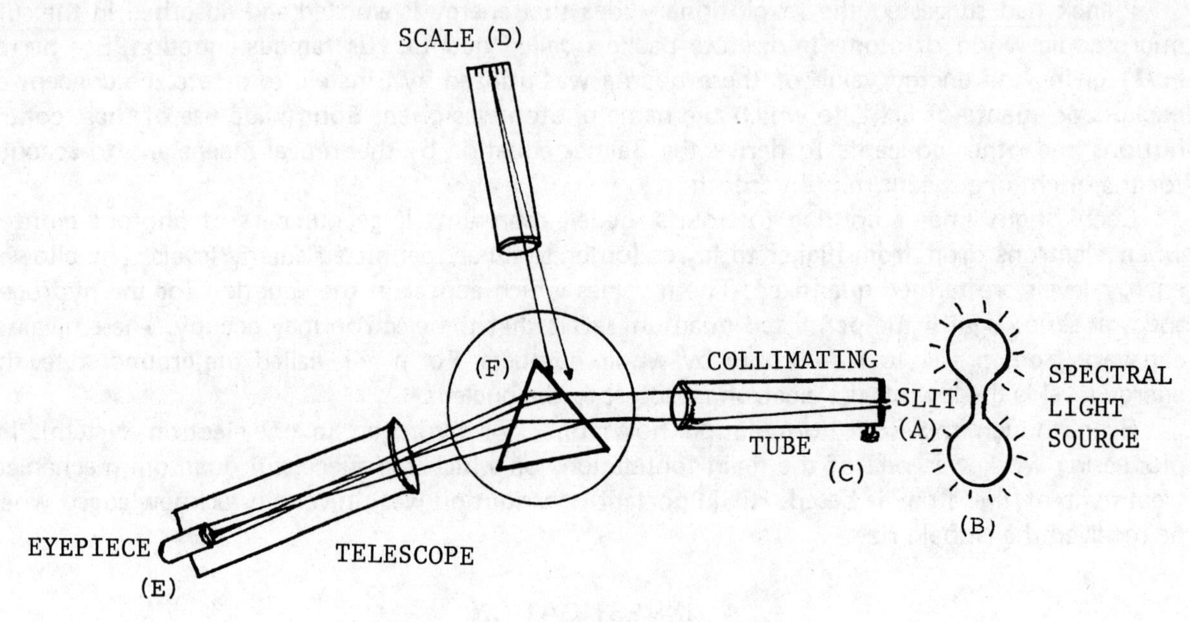

Figure 11. The visual spectrometer.

d. The bottom tube contains the viewing eye piece (E) which can be moved in and out to focus the observed spectrum clearly. The large knurled knob underneath this tube may be rotated so that the viewer may observe various parts of the spectrum. In viewing a spectrum it is wise to move from the left end of the spectrum continuously to the right while taking readings to an estimated tenth place. The reading on each line should be taken twice.

e. The center top cover of the spectrometer can be lifted to observe the quartz prism (F) which disperses the incoming light into separate wavelengths.

II. The Use of the Mercury Spectrum to Calibrate the Scale

a. Plug the tungsten lamp into an electrical outlet and place it behind the tube containing the scale. Cover the lamp with aluminum foil to prevent its light disturbing you while you take spectrum readings.

b. Since mercury arcs are dangerous to the eyes because of the emitted ultra-violet radiation, do not look directly into the arc. The lamp is lined up with the slit (A) (see diagram). Both the lamp and the end of the collimating tube containing the slit are covered with a large piece of aluminum foil to prevent stray radiation.

Plug the mercury arc into an electrical outlet, turn on the switch and depress the button on the transformer for a period of 30 to 60 seconds to light the arc.

Permit the lamp to warm up for 5 minutes.

c. To calibrate the prism spectroscope, rotate the spectrometer scale knob toward the left end of the scale until you see the red lines associated with the spectrum. Adjust the eye piece for clarity. Sharpen the lines with the slit adjusting screw, but, of course, do not close the slit to the point where the spectrum disappears.

d. Use the wavelengths for mercury given on page 55 and identify the lines you observe. Readings should be taken from left to right (i.e., from the longer wavelength, red end of the spectrum, to the shorter wavelengths in the blue end). There are nine lines to be identified but it is possible that you will see many more than these in going through the spectrum. The longer the mercury arc is on, the more lines will be visible. These are usually due to emissions from impurities.

However, by noting the intensities and the relative positions of the lines it will not be difficult to match up the brightest lines with the accepted wavelengths.

Record the scale reading, color, and intensity of each line on the data sheet. When finished, turn off the mercury arc.

III. Determination of the Visible Wavelengths of the Hydrogen Spectrum

a. Replace the mercury arc with the hydrogen discharge tube. Align the lamp with the spectrometer as before, with the narrow section of the discharge tube parallel to the slit. Plug the discharge tube into a transformer and turn on the switch. The warm-up period is about 3 minutes.

b. Without changing the slit width take readings of the *three* lines which you can observe. Record your observations.

c. Turn off the hydrogen lamp.

d. Given the following *mercury arc wavelengths* in nonometer, assign them to your spectral scale readings: 623.4, 615.2, 579.0, 577.0, 546.1, 502.5, 435.8, 407.8, 404.7.

e. Plot a graph of scale reading versus wavelength but be sure to use graph paper in which scale reading can be defined to ±0.1 and that wavelengths are readable to ± 1 nm. Start the x-axis at 350 nm, the beginning of the visible range. You may have to tape graph paper together in order to do this. This is your calibration curve for the spectrometer.

A typical calibration curve for scale readings of the spectrometer versus wavelength looks like this:

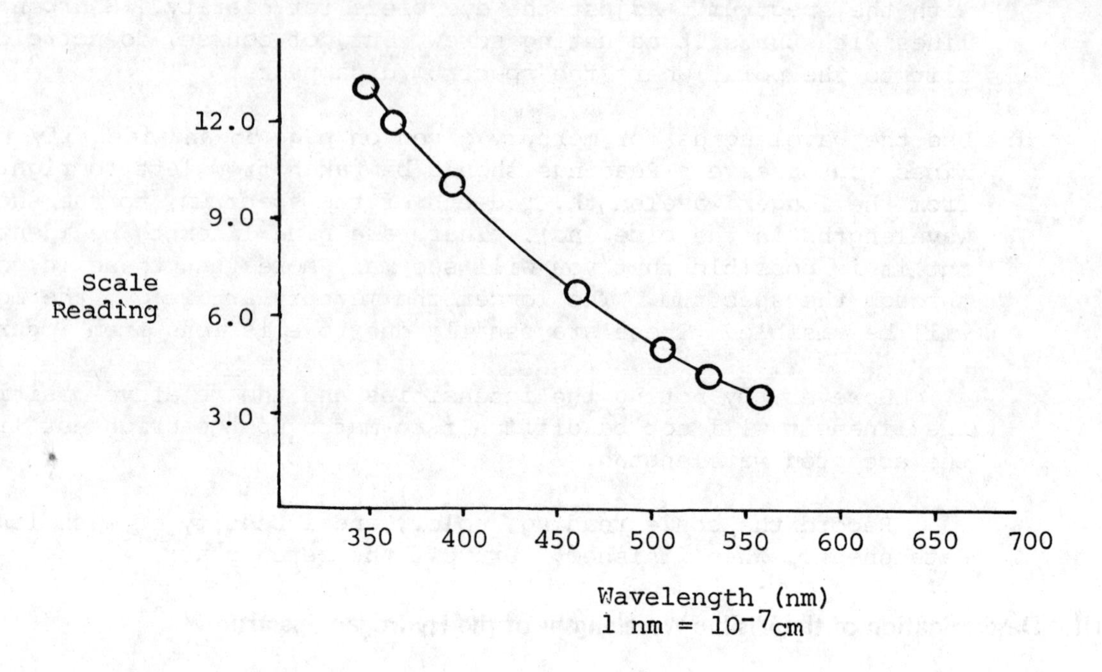

f. Using your calibration curve determine the wavelengths of your Balmer series lines. Record on data page. (III b)

IV. Calculation of the Rydberg Constant

a. Calculate the reciprocal of the wavelength $\left(\frac{1}{\lambda}\right)$ for these three lines.

b. The general form of the spectral series equation is:

$$\frac{1}{\lambda} = R\left(\frac{1}{n_i^2} - \frac{1}{n_o^2}\right)$$

Since n_i is a constant for a particular set of lines, it is convenient to rewrite the equation as:

$$\frac{1}{\lambda} = -R\left(\frac{1}{n_o^2}\right) + \frac{R}{n_i^2}$$

Written in this way it is more apparent that we have the equation of a straight line. (Recall that the equation for a straight line is $y = mx+b$). $1/\lambda$ varies linearly with values of $1/n_o^2$, $-R$ is the slope and R/n_i^2 is the y intercept.

Assign an arbitrary integer value (n_o) to each line of the hydrogen spectrum. The line corresponding to the longest wavelength should be given the smallest n_o value and each line following is then assigned, consecutively, a value of n_o+1 and n_o+2. If you start with n = 2 then the lines following correspond to an n value of 3 and 4 respectively.

In the Balmer series, the value of n_i is 2. Therefore, values of n_o must be $n_i + 1$, $n_i + 2$,.... Prepare a careful plot of $\frac{1}{\lambda}$ on the y-axis versus $\frac{1}{n_o}$ on the x-axis.

Determine the slope of the line graphed in (b). The value of the slope is the negative of the Rydberg constant (-R).

V. Calculation of the Energy Level Differences (ΔE) From Observed Wavelengths

Bohr theory assumes that each line results from the emission of a photon when an electron falls from a higher energy level to a lower one.

Therefore, each photon represents the difference (ΔE) between two energy states. Calculate the three ΔE values from the wavelength of your observed lines.

$$\Delta E = \frac{hc}{\lambda} \text{ (in joules)}$$

Remember to convert your wavelengths to meters. Enter the results of your calculations on the data page.

VI. Calculation of Theoretical Values for Energy Levels Using the Bohr Equation

$$E = \frac{-2\pi^2 m e^4 z^2}{n^2 h^2}$$

where $m = 9.10 \times 10^{-31}$ kg $z = 1$ (the atomic number of hydrogen)

$e = -1.6 \times 10^{-19}$ coulombs $h = 6.62 \times 10^{-34}$ joule sec

calculate theoretical values for the first five energy levels of the electron in a hydrogen atom (i.e., for n = 1, 2, 3, 4, 5).

You can now calculate the theoretical differences in energy between the levels n = 1, 2, 3, 4, 5. ($\Delta E_{i-o} = E_i - E_o$, etc.)

Record the results on your data page.

[illegible]

[illegible]

[illegible]

IV. Calculation of [illegible] Different [illegible] from Observed Wavelengths

[illegible]

[illegible]

[illegible]

[illegible]

VI. [illegible] of Theoretical [illegible] Levels Using [illegible]

[illegible]

[illegible]

NAME____SECTION______________DATE______________

BALMER SERIES FOR HYDROGEN

DATA

II. d. Calibration of the Spectrometer Scale for the Mercury Spectrum:

Color	Scale Reading	λ(nm)
		6234
		6152
		5790
		5770
		5461
		5025
		4358
		4078
		4047

III. b. Hydrogen Spectrum:

Color	Scale Reading	λ(nm) Read from calibration curve

IV. Determination of the Rydberg constant:

color	λ (see c)	$\frac{1}{\lambda}$ (see d)	n_o (see e)	$1/n_o^2$	n_i (see h)

Experimental value for the Rydberg constant from the slope of the graph. (Show method of calculations)

Accepted value of the Rydberg constant ____________________
Percent error. (Show calculations)

V. Calculation for Energy Level Differences, ΔE, from observed wavelengths: (Show method of calculation)

λ	$\frac{1}{\lambda}$	ΔE

VI. Calculated values of the energy levels for the electron in the hydrogen at

a. for n = 1 E =

for n = 2 E =

for n = 3 E =

for n = 4 E =

for n = 5 E =

b. Calculation of theoretical differences (ΔE)

THOUGHT

1.a. The Lyman Series of wavelengths orignates when an excited electron falls to the n = 1 level. Calculate the wavelengths of the first 3 lines of this series for the hydrogen atom.

b. In what region of the electromagnetic spectrum do these lines lie?

NAME______________________________________SECTION____________DATE____________

Experiment 4

PERCENT COMPOSITION OF A COMPOUND

PRELIMINARY QUESTIONS

1. The chemical combination of copper and oxygen appears to produce a product containing large variations in percentages of copper and oxygen by weight. Is this an exception to the law of definite proportions? Can you propose a possible explanation for this apparent anomaly?

2. A sample of table salt was found by multiple analyses to contain 2.36 g of sodium and 3.64 g of chlorine. Calculate the percent composition of sodium and chlorine. Using the accepted atomic weights, find the empirical formula.

3. One pound of sulfur combines with one pound of oxygen to produce a gas. What is the percent by weight composition of this gas?

4. Write balanced chemical equations for the following reactions:

   a) for the reaction of magnesium with oxygen of the air to form magnesium oxide.

   b) for the reaction of magnesium with nitrogen of the air to yield magnesium nitride.

   c) for the reaction of magnesium nitride with water.

d) for the reaction of magnesium oxide with water.

e) for the heating of magnesium hydroxide.

Experiment 4
ERCENT COMPOSITION OF A COMPOUND

IDEAS

In the early nineteenth century, two French scientists argued about the composition of chemical compounds. C. Berthollet asserted that the composition of a compound depended upon the proportions of the substances reacting or the conditions of the reaction. On the other hand, J. Proust insisted that each individual chemical compound was always made up of the same elements, which were in a definite proportion to each other by weight. All samples of this compound, from any source, always have exactly the same composition by weight. Let's test these hypotheses.

We are going to produce a compound, magnesium oxide, by heating magnesium in the presence of air.

There are four aspects to this experiment; first: the chemical reactions involved in changing magnesium to magnesium oxide by burning it in the air; second: an analysis of the class results in terms of percent magnesium in magnesium oxide to determine the average value, most frequent value, the range of values and the standard deviation; third: the formula of magnesium oxide based on the experimental results; and fourth: to decide, based on class results, who was right -- Berthollet or Proust.

The discussion on how to analyze the data can be found in the introductory discussion, pp. 1-7.

Outline of Reactions

When magnesium, a very reactive metal, is burned in air, it reacts not only with oxygen to form magnesium oxide, but also with nitrogen to form magnesium nitride.

$$2Mg + O_2 \longrightarrow 2MgO$$

$$3Mg + N_2 \longrightarrow Mg_3N_2$$

Since the object of the experiment is to find out how much magnesium oxide forms from a known weight of magnesium, it is necessary to convert all the magnesium nitride to magnesium oxide. The following procedure brings this about.

Water is added to the mixture of magnesium oxide and magnesium nitride:

$$MgO + H_2O \longrightarrow Mg(OH)_2$$

$$Mg_3N_2 + 6H_2O \longrightarrow 3Mg(OH)_2 + 2NH_3$$

The ammonia gas is lost to the atmosphere; the product remaining in the crucible is only magnesium hydroxide and some excess water. The final step is to heat this magnesium hydroxide to drive off the water:

$$Mg(OH)_2 \longrightarrow MgO + H_2O$$

The empirical formula can be calculated by 1) finding the weight in grams of both magnesium and oxygen which combined; 2) calculating the number of moles of each represented by these weights; 3) finding the molar ratio of magnesium to oxygen; and 4) then determining the formula by finding the smallest whole numbers which best represents this ratio.

Here is a sample calculation using a different oxide:

2.00 g of iron reacted to form 2.86 g of iron oxide.

The number of grams of oxygen is 2.86 – 2.00 = 0.86 g

The number of moles of iron is $\frac{2.00\ g}{56\ g/mole} = 0.036$ moles iron

The number of moles of oxygen is $\frac{0.86\ g}{16\ g/mole} = 0.054$ moles oxygen

Molar ratio $\frac{O}{Fe} = \frac{0.054}{0.036} = \frac{1.5}{1}$

Smallest whole number ratio: $\frac{(1.5)(2)}{(1)\ (2)} = \frac{3}{2}$

Formula: Fe_2O_3

On the basis of your own experiment, is it possible to evaluate the ideas of Proust and Berthollet? By comparing the results of all the students in the class, can you evaluate the alternative hypotheses? Obviously, it is important for you to follow the instructions carefully for making magnesium oxide for your results to be meaningful.

INVESTIGATION

Purpose: To find percent composition and empirical formula of magnesium oxide; to evaluate experimental data.

Equipment:

1 crucible and cover
1 Bunsen burner
1 tripod
1 clay triange
1 glass stirring rod
crucible tongs
red litmus paper

Chemicals: 2 40-cm strips of magnesium ribbon

Procedure:

Heat, cool, and weigh precisely a clean crucible and cover. Take a 40-cm strip of magnesium, cut it into two roughly equal pieces, and roll each piece into a flat, loose coil. Place both coils in the bottom of the crucible. Weigh the crucible containing the magnesium ribbon, and cover. Support the covered crucible and contents on a clay triangle. Adjust the cover so that the crucible is slightly open to permit some air to enter. *Smoke is evidence that some product is being lost. Heat slowly. Do not allow the reaction to smoke.* Use the crucible tongs to adjust the cover so that the reaction proceeds without the loss of material from the crucible. Continue heating until the magnesium no longer glows brightly, plus another 5 minutes.

Remove the cover and heat strongly for about 5 minutes. Allow to cool. With your glass stirring rod, carefully crush the product in the crucible into a powder. Tap any powder sticking to the stirring rod into the crucible.

Add about 0.5 ml (10 drops) of deionized water dropwise, rinsing any specks of solid off the glass stirring rod. Warm slowly to allow the water to evaporate without spattering. Detect the odor of ammonia which is produced from magnesium nitride. The presence of ammonia can be confirmed by holding moist red litmus paper over the warmed mixture and observing the color change to blue. Partially cover the crucible and heat strongly for at least 10 minutes. Strong heating is necessary to convert magnesium hydroxide to magnesium oxide. The bottom of the crucible should be red hot. Allow to cool to room temperature and weigh.

Repeat the experiment for a second value.

Purpose:

Procedure:

Name_______________________________________SECTION__________DATE______________

PERCENT COMPOSITION OF A COMPOUND

DATA

1. Why is strong heating necessary at the end of the experiment?

		Trial 1	Trial 2
a.	Weight of crucible, cover and magnesium	______	______
b.	Weight of crucible and cover	______	______
c.	Weight of magnesium	______	______
d.	Weight of crucible, cover and magnesium oxide	______	______
e.	Weight of crucible, cover	______	______
f.	Weight of magnesium oxide	______	______
g.	Experimental percent magnesium in magnesium oxide (show method of calculations clearly). Report to the proper number of significant figures.	______	______

Using accepted atomic weights, calculate:

h.	Number of moles of magnesium atoms in (c)	______	______
i.	Number of moles of oxygen atoms combined with the magnesium	______	______
j.	Formula of magnesium oxide from your experimental values	______	______
k.	Theoretical percentage of magnesium in magnesium oxide	______	______
l.	Percent error (show method of calculation)	______	______

As soon as you have completed the calculations, enter your values on the class data sheet.

THOUGHT

1. a. Compute the average percent magnesium for all the values of the class. ______

b. What is the range of values? ______

c. Calculate the standard deviation.

d. Examine the class data. Count the number of times each value for the percent magnesium appears. On graph paper, plot the numbers (frequency) on the y-axis against the percent magnesium on the x-axis.

e. On the graph mark off the most frequent value, the average value, and the accepted value.

f. Are these values (in e) the same when considered to 2 significant figures?

A graphical picture of a large collection of data can often give clues concerning the validity of the experimental procedure. For example, if there is great disagreement among the values, it probably means that some variable has not been controlled. Or if the curve is skewed, it can mean that there is a systematic error. Consider the following questions and answer them with respect to the effect of each on the calculated percent magnesium in the sample, i.e. high, low or no effect on the percent magnesium in the oxide. Some questions may require information of the kind that can be found in the *Handbook of Chemistry and Physics*.

2. a. Is there more than one compound of magnesium and oxygen?____________
Are all these compounds likely to form under the conditions of the experiment? Explain.

b. Can the magnesium evaporate under the conditions of this experiment?

c. If all the magnesium does not burn, what will be the effect on the calculated percent magnesium in the oxide?

d. The reaction smokes briefly. What is the smoke due to?
What is the effect on the calculated percent magnesium in the oxide?

e. All the magnesium nitride is not converted ultimately to magnesium oxide. What is the effect on the calculated percent magnesium in the oxide?

f. All the water is not removed from the end product. What is the effect on the calculated percent magnesium in the oxide?

NAME______________________ SECTION__________ DATE__________

g. Some of the product is lost by spattering. What effect will this have on the calculated value for the percent magnesium in the oxide?

h. Using the class results as the basis for your answer, can you decide who was right, Berthollet or Proust? Explain.

3. a. Can one determine the *atomic weights* from knowing only the percent composition of the elements in a compound?

b. Can one determine the *formula* of the compound knowing only percent composition?

c. If you know only the percent composition of magnesium oxide and atomic weight of oxygen, 16 amu, can you calculate

i) the atomic weight of magnesium?

ii) the molecular weight of the oxide?

NAME_______________ SECTION_______________ DATE_______________

Experiment 5

DETERMINATION OF APPROXIMATE ATOMIC WEIGHT (DULONG AND PETIT)

PRELIMINARY QUESTIONS

1. Define: (a) calorie, (b) specific heat.

2. a. When one kg of water at 18°C is mixed with one kg of mercury at 30°C, the resulting temperature of the mixture is found to be 18.4°C. Compute the specific heat of mercury. (Assume s_{H_2O} is 1.0 cal/g°C.)

   b. How does the specific heat of mercury compare with that of water?

3. According to the definition of the calorie as a quantity of heat, exactly what variables must be known to determine heat loss or gain?

4. State the rule of Dulong and Petit.

5. An element A has a specific heat of 0.028 cal/(g) (°C). 10.000 g of the element reacted with oxygen to form an oxide weighing 11.142 g. What is the exact atomic weight of the element?

6. Calculate the percent error of the atomic weight obtained by using the rule of Dulong and Petit.

7. Although only the approximate atomic weight is determined by the method of Dulong and Petit the approximation does not affect the usefulness of this rule. Explain.

Experiment 5

DETERMINATION OF APPROXIMATE ATOMIC WEIGHT (DULONG AND PETIT)

IDEAS

In 1819 P. Dulong and A. Petit discovered a quantitative relationship between the specific heat of an element and its atomic weight:

$$(\text{specific heat})\ (\text{atomic weight}) = 6.3 \qquad \text{This product is called atomic heat.}$$

$$\frac{(\text{calories})}{(\text{g})(^\circ\text{C})}\ \frac{(\text{grams})}{(\text{mole})} = 6.3\ \frac{\text{calories}}{(^\circ\text{C})(\text{mole})}$$

Called the Law of Dulong and Petit, this is an excellent example of an empirical law, that had no theoretical foundation at the time it was formulated. It provides approximate values for the atomic weight of metals above aluminum.

Specific heat is defined as the amount of heat (calories) required to raise the temperature of one gram of a substance one degree Celsius. A calorie has been defined as the amount of heat necessary to raise the temperature of one gram of water one degree Celsius.

Specific heat is a measure of the heat capacity of a substance, or its ability to absorb heat, or its "appetite" for heat. If equal masses of water and mercury at the same temperature in identical containers are placed in the direct sun on a summer day, the temperature of the mercury will be raised much more than the temperature of the water after any specific time interval. (Compare iron and wood.) The water therefore has a greater heat capacity, or "appetite" for heat, for it requires more heat than mercury to achieve the same rise in temperature.

Joseph Black, in the eighteenth century, mixed a weighed quantity of water at a measured temperature with a weighed quantity of some other substance at a measured but higher temperature and then determined the temperature of the mixture. This method of mixtures was designed to compare the heat capacities of various substances even before an adequate theory of heat was developed. According to the caloric theory of heat accepted at that time, heat was a fluid which was conserved when it flowed from one object to another. If two substances with different temperatures were mixed, the heat gained by one was therefore equal to the heat lost by the other. Although the caloric theory of heat has been replaced by the kinetic molecular theory, the idea of conservation of heat in this method of mixtures is still valid. It can be expressed in the following way:

$$\text{heat gained} = \text{heat lost}$$

The quantity of heat, in terms of calories, which is gained or lost when heat is transferred can be computed by

$$(m)(t_f - t_i)(s) = (m)(t_i - y_f)(s)$$

m = mass in grams

$t_f - t_i$ or $t_i - t_f$ is the change in temperature °C.

t_f = final temperature °C.

t_i = initial temperature °C.

s = specific heat in calories/gram °C.

Precise atomic weights of metals* can be obtained through the careful analysis of compounds of the metal with oxygen, the calculation of the equivalent weight of the element, and the application of the law of Dulong and Petit. Equivalent weight is defined as the number of grams of an element which combines with 8.000 grams of oxygen. An example will make this clear.

The following experimental information is obtained about a certain metallic element, M:

the specific heat of the metallic element M is 0.11 cal/g°C.

2.792 g of this element combined with oxygen to form 3.992 g of an oxide.

The approximate atomic weight is calculated using the law of Dulong and Petit.

$$\frac{(6.3 \text{ cal})(\text{g °C})}{(\text{°C})(\text{mole})(.11 \text{ cal})} = \frac{57 \text{ g}}{\text{mole}}$$

The exact equivalent weight can be calculated as follows:

3.992 g oxide - 2.792 g metal = 1.200 g oxygen

$$\frac{\text{g of M}}{\text{equivalent}} = \frac{(8.000 \text{ g O})(2.7924 \text{ g M})}{(\text{equivalent})(1.200 \text{ g O})} = \frac{18.613 \text{ g M}}{\text{equivalent}}$$

The approximate atomic weight of 57 is about three times the exact equivalent weight of 18.613. Therefore an exact atomic weight can be obtained by multiplying the equivalent weight by 3:

$$\frac{(18.613 \text{ g of M})\ (3 \text{ equivalents})}{(\text{equivalent})\ (\text{mole})} = \frac{55.840 \text{ g of M}}{\text{mole}}$$

In the twentieth century, Albert Einstein's study of the specific heat of solids led to his suggestion (1907) that oscillating atoms within solids absorb heat energy in discrete packets rather than continuously. This was an historic extension of Max Planck's quantum theory.

*This method applies to metallic elements which have atomic weights larger than aluminum. There is also a slight error introduced by using an equivalent weight which is based on the old atomic weight standard for oxygen of 16.000 amu.

INVESTIGATION

Purpose: To determine the specific heat of an element; to apply the law of Dulong and Petit to find the approximate atomic weight of an element.

Equipment:

1 400-ml beaker
1 25-cm test tube with grooved cork
2 thermometers (110°C)
1 50-ml graduated cylinder
1 styrofoam cup, used as a calorimeter

Chemicals: metallic element, in small pieces: lead, Pb; zinc, Zn; iron, Fe; nickel, Ni; etc.

Procedure:

Add 50 ml of water, measured with a graduated cylinder, to your styrofoam cup (calorimeter), and insert a thermometer into the water. Weigh precisely 30 to 35 g of a *dry* metallic element into your test tube. Bore a hole in the cork to fit your thermometer. With your triangular file, groove a slot vertically in the cork. When you assemble the apparatus, insert the thermometer carefully into the test tube. Shaking or tilting the tube will help to position the thermometer properly. Support the test tube vertically in the beaker with of the beaker. Add water to the beaker so that its level is about 2 cm above the level of the metal in the test tube.

Boil the water to heat the element to nearly 100°C. When the temperature of the metal is near or at 100°C, record the temperature of the water in the calorimeter. Pour the metal rapidly into the water in the calorimeter and stir. Record the maximum temperature of metal and water. This temperature is used to calculate the heat loss of the metal as well as the heat gain of the water. With these data calculate the specific heat and the approximate atomic weight of the element. Repeat the experiment to obtain reproducible values.

INVESTIGATION

NAME______________________________ SECTION__________ DATE__________

DETERMINATION OF APPROXIMATE ATOMIC WEIGHT (DULONG AND PETIT)

DATA

	metal 1	metal 2
a. weight of test tube, beaker and element	_____g	_____g
b. weight of test tube and beaker	_____g	_____g
c. weight of element	_____g	_____g
d. temperature of water in the calorimeter before mixing (t_i, H_2O)	_____°C	_____°C
e. temperature of the hot element before mixing (t_i, metal)	_____°C	_____°C
f. final temperature of mixture (t_f)	_____°C	_____°C
g. $t_f - t_i$ for H_2O	_____°C	_____°C
h. $t_i - t_f$ for the metal	_____°C	_____°C

i. Calculated specific heat of the element. Show method of calculation.

metal 1 _____ cal/g°C

metal 1 _____ cal/g°C

j. Calculated atomic weight of the element. Show calculations.

metal 1 _____ g/mole

metal 2 _____ g/mole

k. Accepted atomic weight, if known

metal 1 _____ g/mole

metal 2 _____ g/mole

l. Percent error of the known element. Show calculations.

metal 1 _____

metal 2 _____

THOUGHT

1. Why must the hot metallic element be dry before it is poured into the water in the calorimeter?

2. Why was it necessary to cut a groove in the cork?

3. Why can you disregard the heat gained by the calorimeter?

4. What are the sources of error in your experimental operations?

5. Give two reasons why you can report your results with a precision of only two significant figures.

NAME________________________SECTION__________DATE__________

Experiment 6

ANALYSIS OF A COPPER-NICKEL ALLOY

PRELIMINARY QUESTIONS

1. Define:

   a) purity

   b) solubility

   c) decantation

   d) ion

2. What is a qualitative chemical test?

3. Which is more effective: to wash an insoluble precipitate with 15 ml of water once or to wash the precipitate with 3 ml of water 5 times?

4. Write balanced equations for the following reactions:

   a) Copper and nitric acid yield copper (II) nitrate, nitrogen dioxide and water.

   b) Nickel and nitric acid yield nickel (II) nitrate, nitrogen dioxide and water.

   c) Zinc and copper chloride yield zinc chloride and copper.

   d) The reaction of nitrogen dioxide with water yields nitric acid and nitric oxide (NO).

5. Write formulas and indicate charge for the following complex ions:

   a) copper (II) tetraammine

   b) nickel (II) hexaammine

   c) zinc (II) tetraammine

   d) hexacyanoferrate (II)

6. Write formulas for the following compounds:

   a) nickel dimethylglyoxime

   b) copper (II) hexacyanoferrate (II)

Experiment 6

ANALYSIS OF A COPPER-NICKEL ALLOY

IDEAS

Today you will separate copper from nickel quantitatively in order to find the percent composition of the alloy. The making of alloys was a practical art even in ancient times. Objects thousands of years old made of alloys have been discovered in archaelogical investigations, and recipes for making alloys have been found in Egyptian manuscripts (about 3 AD). Articles estimated to be as old as 4000 years, made of the alloy bronze and containing varying percentages of copper and tin, were discovered by the pioneering archaelogist Schliemann in the ruins of Troy and in other excavations.

Pliny, the Elder, the Roman administrator and scholar of the first century AD, who was a prolific writer on many subjects, described methods of separating impurities from gold ore. Although the base metals were easily removed, it was more difficult to remove silver. Gold objects from antiquity have been found with varying percentages of silver. Ancient tests for purity of gold were color and specific gravity. In the famous legend about Archimedes, he determined whether a crown was made of pure gold by the specific gravity method which he is said to have invented. Gold is more dense than silver. Archimedes found that an object immersed in water will displace a volume of water equal to its volume. He compared the volume of a crown of pure gold and one of the same mass made of an alloy of gold and silver.

Attempts to make imitation gold, silver, or alloys of the two from cheaper materials have been recorded in an Egyptian papyrus (about 3 AD). Although these manuscripts indicate that practitioners of these arts clearly understood that the recipes produced imitations, other writers of the period believed in the possibility of actually converting or "transmuting" base metals into gold or silver -- the art of alchemy. The roots of this apparently strange quest have been attributed in part to the belief of both Plato and Aristotle that all matter was made up of the four elements, air, water, fire, and earth, which could be converted from one to another under suitable conditions. Alchemical activity continued through the Middle Ages. Because many honestly believed in the possibility of converting cheaper metals to "nobler" ones, the opportunity to take advantage of the gullible proved irresistible to those who are always looking for easy money and the alchemical profession became saturated with charlatans. The profession of alchemy, which was the only chemistry at that time, thus became disreputable. A great improvement in the status of the field was made in the sixteenth century as a result of the energetic efforts of Paracelsus (1493-1541) who vigorously asserted that the alchemists or chemists should devote their efforts to preparing remedies rather than to preparing gold, that the processes going on in the body resemble "alchemical" processes, and that physicians should rely on their own observations rather than blindly follow the advice of the ancient authorities.

Because of the work of Paracelsus, many physicians and chemists recognized the productive relationship between their fields, and so many talented individuals took a greater interest in an area that had previously been scorned.

Today, the alchemist's dream is a reality. It is possible to transmute one element into another by nuclear reactions, which release tremendous amounts of energy.

If two or more metals are melted together and allowed to solidify, the product is called an alloy. Most alloys are solid solutions. Alloys are useful in that specific desired properties can be imparted. For instance, silver and copper have been blended to produce the alloy coinage silver. Silver metal is soft, but the alloy of silver and copper is hard enough to be used for coins. This alloy is no longer used in the United States for silver currency. Instead, copper and nickel are blended to form a tough silver-appearing alloy for the exterior skin. The interior of the coin is copper and is visible along the milled edge. The amounts of each metal used simulates not only the appearance of the "silver" coins, but also approximates the density of the older coins.

Analysis of alloys can be made both by physical and chemical methods. Physical separations are those which are accomplished by bringing about changes in state of the materials, or by taking advantage of other differences in physical properties, such as solubility or crystallization of the materials. Chemical separations are those in which the substance is changed into one or more new substances with different properties. In this experiment, the separations are accomplished first by treating the alloy with reagents to bring about chemical changes. The resulting products are easily separated by the physical methods of filtering or decanting.

INVESTIGATION

Part A

Purpose: To separate quantitatively copper from nickel in an alloy; to determine the percent copper in the alloy.

Equipment:

- 1 150-ml beaker
- 1 50-ml graduated cylinder
- 1 glass stirring rod
- 1 short stem funnel and filter paper
- 1 100-ml graduated cylinder
- 9 10-cm test tubes
- 1 watch glass
- 1 tripod
- 1 asbestos wire gauze
- 1 hot plate

Chemicals:

- nickel-copper alloy, Ni/Cu
- nitric acid, HNO_3, conc
- hydrochloric acid, 6 M
- mossy zinc, Zn
- ammonia, NH_3, 6 M
- potassium hexacyanoferrate (II) $K_4Fe(CN)_6$, 0.1 M
- nickel chloride, $NiCl_2$, 0.1 M
- copper chloride, $CuCl_2$, 0.1 M
- zinc chloride, $ZnCl_2$, 0.1 M
- dimethylglyoxime, 1%, in ethanol
- methanol
- red and blue litmus paper

Outline of Procedure

A nickel-copper alloy is dissolved in concentrated nitric acid (HNO_3) to yield copper (II) nitrate, $Cu(NO_3)_2$, nickel nitrate, $Ni(NO_3)_2$, and nitrogen dioxide, NO_2, a brown gas. The resulting solution is heated with 6 M HCl almost to dryness to remove the nitrate groups and to convert the salts to chlorides. Zinc metal added to an acid solution of the two salts, replaces the copper ion from the copper chloride, $CuCl_2$, precipitating metallic copper, and releasing hydrogen gas, H_2. The reaction is quantitative in that all of the copper in the solution can be precipitated by this method. By knowing the mass of the alloy and the mass of the precipitated copper metal, it is possible to calculate the percent copper in the alloy.*

Procedure:

Wear Safety Glasses. Use Hoods.

Isolation of Copper

I Weigh precisely a 150-ml beaker. Add to the beaker about 2 grams of the nickel-copper alloy and weigh them precisely. Then place the beaker in the hood and add 10 ml concentrated nitric acid. Observe the copious evolution of brown gas. If the reaction appears to stop before the alloy has completely dissolved, warm slightly or add a few milliliters of concentrated nitric acid. When the alloy has completely dissolved, and the reaction is over, add 20 ml 6 M hydrochloric acid, and heat to boiling. Evaporate nearly to dryness. Cool slightly and repeat with a second increment of 20 ml 6 M hydrochloric acid. Again evaporate nearly to dryness.**

Dissolve the nearly dry salts in 25 ml 6 M hydrochloric acid, and transfer the solution to a 100-ml graduated cylinder. Rinse the beaker with deionized water and add the rinsings to the solution in the cylinder. Repeat this procedure until the salt has been quantitatively transferred from the beaker to the cylinder. Fill the cylinder to the 100-ml mark with deionized water.** Mix the contents thoroughly by transferring the entire solution to a clean *dry* 150-ml beaker. The solution should be a homogeneous green color.

---

* The nickel can also be precipitated quantitatively as nickel dimethylglyoxime, a red insoluble precipitate. The amount of nickel in the alloy can be determined from the quantity of this precipitate which is formed.

$$Ni^{2+} + 2C_4H_8N_2O_2 \rightarrow Ni(C_4H_7N_2O_2)_2 + 2H^+$$

Nickel dimethylglyoxime has the structure:

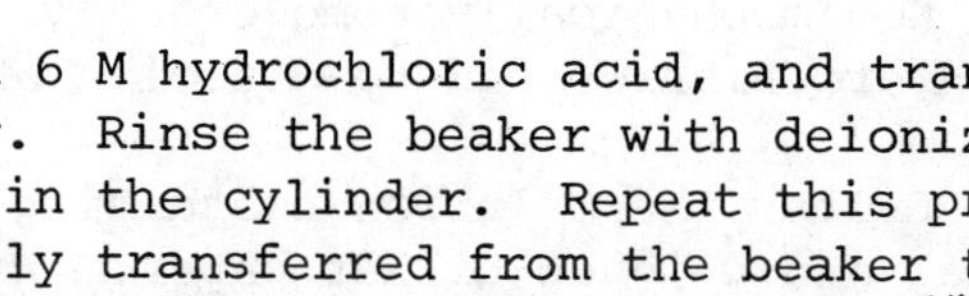

```
           O—H ··O
CH3—C = N/         \N = C—CH3
    |        Ni        |
CH3—C = N\         /N = C—CH3
           O··H—O
```

** If the experiment is to be completed in a following laboratory period, there are two stop convenient stopping points indicated by **. If you stop here, cover your solution and store in your desk. If you stop at the second point, place your solution in a clean dry Erlenmeyer flask closed with a rubber stopper. Proceed with the tests for the ions described in part B.

II. Using a graduated cylinder, transfer 25.0 ml of this solution to a 150-ml beaker.

Assuming your sample of alloy to be pure copper, calculate the theoretical amount of zinc required to precipitate all of the copper. Use your equation for the displacement of copper ion by zinc metal as the basis for the calculation. However, since some of the zinc will be consumed by hydrochloric acid to produce hydrogen the amount used up must be calculated and added to the amount replacing the copper. Add the sum total of these to the container in two steps, by adding half of the total first, and the remainder in small increments after the reaction has slowed down.

Hydrogen is evolved rapidly so keep your reaction *away from flame* and in a well ventilated area.

Notice the color changes of the solution as the reaction proceeds.

When the reaction appears to be complete, i.e., all copper ions have been precipitated as metallic copper, test the solution chemically for the presence of copper ions.* If any blue color is observed, add a small amount of zinc and test the solution again in 10 minutes. Continue in this way until the copper ion test is negative.

Now mechanically remove any large identifiable pieces of zinc and add 10 ml 6 M hydrochloric acid to dissolve any remaining zinc.

Filter the solution containing the metallic copper using the short stem funnel and weighed filter paper. Wash the precipitate of copper with several small increments of distilled water until the filtrate, when tested for the presence of nickel ions (using dimethylglyoxime and a few drops of 6 M ammonia) and zinc ions (using potassium hexacyanoferrate (II)) gives a negative result.

Follow this washing with three 10 ml increments of methanol to dry the metallic copper. Spread the filter paper on a weighed watch glass and allow the precipitate to air dry, or place the filter paper and watch glass with the copper in an oven set at 65°C for one and one half hours. In either case, test the precipitate for dryness by weighing to constant weight.

---

* See part B; Tests For The Ions

NAME______________________________ SECTION____________ DATE__________

ANALYSIS OF A COPPER-NICKEL ALLOY

Part A

DATA

. Wt. of 150-ml beaker and Ni-Cu alloy __________ g
. Wt. of 150-ml beaker __________ g
. Wt. of Ni-Cu alloy __________ g

. Wt. of copper, watch glass, and filter paper __________ g
. Wt. of watch glass __________ g
. Wt. of filter paper __________ g
. Wt. of copper recovered __________ g
. Wt. of copper in the entire sample of alloy __________ g

Percent of copper in the alloy __________

II. i. How much of your alloy is present in the solution after you have transferred 25 ml of the dissolved alloy to a beaker?

ii. Write an equation for the displacement of copper ion by zinc metal.

iii. a. Calculate theoretical amount of zinc required to precipitate all the copper from 25 ml of the alloy solution. (neglect the acid present)

____

b. Calculate the theoretical amount of zinc that reacts with 25 ml 6 M hydrochloric acid.

____

c. How much zinc would be required to react with only ¼ of the prepared solution of alloy? (i.e. solution containing both acid and alloy)

iv. How do you know that copper ions are being removed from the solution?

v. What is the chemical test for copper ions?

vi. How can you tell when all the zinc has dissolved?

vii. Write a balanced equation for the chemical reaction of zinc with hydrochloric acid.

THOUGHT

1. What metallic impurities could be present in your sample of copper?

2. How would you test the *metal* to determine the presence of these impurities?

3. a. Does your sample of copper have white or green areas of discoloration?

   b. If so what could cause this and how could this have been prevented?

   c. What effect will these salts have upon the calculated percent copper in the alloy?

4. a. What is the color of copper (ii) oxide?

   b. Does your sample of copper show evidence of oxide formation?

   c. What effect would oxidation of the copper have on your determination of percent copper in the alloy?

INVESTIGATION
Part B

Purpose: To study reactions which permit the identification of specific ions (qualitative chemical tests).

Procedure:

Qualitative Tests for the Ions

To develop specific qualitative, sensitive tests for copper, nickel, and zinc ions set up a series of nine 10-cm test tubes, three with 0.1 M copper chloride, three with 0.1 M nickel chloride, and three with 0.1 M zinc chloride. Use only about 1 ml of the 0.1 M reagent for each.

In each of the tests described below, observe color changes of solutions, precipitate formation or disappearance, and colors of precipitates. A solution is perfectly transparent regardless of its color. If there is any cloudiness or turbidity in the solution, a precipitate is present.

1. In the first test of the ions, the addition of 6 M ammonia is carried out in two steps.

   a. To each solution add 1 drop 6 M ammonia. *Be sure to stir each solution thoroughly*. Record results.

   b. Then add 2 ml 6 M ammonia to make the solution distinctly ammonical. Stir well and record the results.

2. In the second test, add a few drops of 1% dimethylglyoxime. Record the results.

3. In the third test, add a few drops of 0.1 M potassium hexacyanoferrate (II) and record the results. Tabulate your data.

NAME______________________________ SECTION____________ DATE____________

ANALYSIS OF A COPPER-NICKEL ALLOY

Part B

DATA

Results of Tests

	Cu^{2+}*	Ni^{2+}*	Zn^{2+}*
1. 6 M ammonia			
a. small amount			
b. excess			
2. 1% DMG**			
3. 0.1 M $K_4Fe(CN)_6$			

* The solutions should be chlorides.

** DMG is an abbreviation for dimethylglyoxime. See footnote with single asterisk on page 71 for a discussion of the dimethylglyoxime reaction. The colored solution that is produced by the reaction of Cu^{2+} and DMG is not the best test for the identification of copper.

1. Balance the following net ionic equations for each of the following reactions:

a. copper ions and small amount of ammonia

$$Cu^{2+} + NH_3 \cdot H_2O \rightarrow Cu(OH)_2 + NH_4^+$$

b. copper hydroxide and excess ammonia

$$Cu(OH)_2 + NH_3 \rightarrow Cu(NH_3)_4^{2+} + OH^-$$

c. nickel ions and small amounts of ammonia

$$Ni^{2+} + NH_3 \cdot H_2O \rightarrow Ni(OH)_2 + NH_4^+$$

d. nickel hydroxide and excess ammonia

$$Ni(OH)_2 + NH_3 \rightarrow Ni(NH_3)_6^{2+} + OH^-$$

e. zinc ions and small amount of ammonia

$$Zn^{2+} + NH_3{\bullet}H_2O \rightarrow Zn(OH)_2 + NH_4^{+}$$

f. Zinc hydroxide and excess ammonia

$$Zn(OH)_2 + NH_3 \rightarrow Zn(NH_3)_4^{2+} + OH^{-}$$

g. nickel ions and dimethylglyoxime

$$Ni^{2+} + C_4H_8N_2O_2 \rightarrow Ni(C_4H_7N_2O_2)_2 + H^{+}$$

h. copper ions and hexacyanoferrate (II) ions

$$Cu^{2+} + Fe(CN)_6^{4-} \rightarrow Cu_2Fe(CN)_6$$

i. nickel ions and hexacyanoferrate (II) ions

$$Ni^{2+} + Fe(CN)_6^{4-} \rightarrow Ni_2Fe(CN)_6$$

j. zinc ions and hexacyanoferrate (II) ions

$$Zn^{2+} + Fe(CN)_6^{4-} \rightarrow Zn_2Fe(CN)_6$$

THOUGHT

A specific, unique test for the presence of an ion should provide *clear*, *unambiguous* evidence that the ion is present. Write balanced net ionic equations to illustrate your answers to the following questions:

1. Based on the evidence of the tests that you have performed

   a. Can you give a specific, unique test for the copper ion?

   b. Can you give a specific, unique test for the zinc ion?

   c. Can you give a specific, unique test for the nickel ion?

NAME______________________________________SECTION____________DATE___________

2. If copper ions and nickel ions were present in the *same* solution, can you identify the presence of each in that solution? Explain by writing balanced net ionic equations.

3. If copper and zinc ions were present in the *same* solution, can you identify the presence of each in that solution? Explain by writing balanced net ionic equations.

4. If nickel and zinc ions were present in the *same* solution, can you determine the presence of each in that solution? Explain by writing balanced net ionic equations.

5. If all three ions, copper, nickel, and zinc, were present in the *same* solution, which of them could you identify by a specific test? Explain.

NAME_______________________________________SECTION____________DATE__________

Experiment 7

MOLAR VOLUME OF A GAS

PRELIMINARY QUESTIONS

1. What is standard temperature and pressure (STP)?

2. Define:

   a) molar volume in words

   b) catalyst

   c) degree Kelvin or absolute temperature

3. What are the equations for Charles's law and Boyle's law?

4. Give two gas law equations:

   a) one expression used when the number of gas molecules remains constant (also called the *combined* gas law)

   b) one expression where the number of gas molecules is also a variable, the general gas law.

5. State Dalton's law of partial pressures.

6. The vapor pressure of water at 23°C is 21 mm Hg. If 100 ml of oxygen is collected over water at 23°C and 730 mm pressure, how much pressure is due to water vapor and how much to the pressure of oxygen?

7. Write a chemical equation for the decomposition by heat of potassium chlorate to form oxygen as one of the products.

8. Will the addition of a small amount of manganese dioxide added to potassium chlorate change the above chemical reaction? Explain.

9. 2.00 g of a mixture of potassium chlorate and potassium chloride, upon heating, liberated 400 ml of oxygen. This was collected over water at 23°C and 745 mm pressure. What percentage of pure $KClO_3$ was contained in the 2.00 g of mixture? Base your calculations on mole relationships.

Experiment 7

MOLAR VOLUME OF A GAS

IDEAS

Molar volume is defined as the volume occupied by one mole of a substance. For all ideal gases the molar volume is the same at constant temperature and pressure. This idea was first postulated by Avogadro when he said that equal volumes of gases under the same conditions of temperature and pressure must contain the same number of molecules. The molar volume of a gas is usually defined with pressure and temperature at standard conditions, 760 mm Hg and 273 °K.

The behavior of gases was investigated by the seventeenth century Irish scientist, Robert Boyle who called attention to the inverse relationship between the volume of a gas and its pressure if the temperature is held constant. In 1662 he published the law which bears his name. It may be written $V = k/P$ (the volume of a gas varies inversely with the pressure).

Boyle, devoting his life to the investigation of natural phenomena, was a pioneer in the field of chemistry, and some of his contributions are discussed in the sections on titrations and qualitative analysis.

Although others had observed that gases expand uniformly upon heating, Jacques Charles (1746-1823) was the first to study this expansion quantitatively. He found that for each degree Celsius rise in temperature, the volume of a gas expands 1/273 of its volume at 0 °C. His interest in gases led Charles to build the first hydrogen balloon when he heard of hot air balloons constructed by the Montgolfier brothers in 1783. Charles himself went up a few times, and balloons that could lift men became the rage. As a result of Charles's improvement of the balloon, men for the first time could be airborne.

It has been found that only a perfect or ideal gas will perform exactly according to Charles's law. Real gases at high pressure or low temperature, and gases which are easily liquified, do not follow this law. It is the object of this experiment to test Charles's law under conditions which approach the ideal and to compare the difference between the experimental value for a real gas (air) and the calculated values for the ideal gas.

Charles's law may be written $V = k'T$ (volume of a given sample of gas varies directly with the absolute temperature at constant pressure).

The laws of Charles and Boyle can be combined:

$$\frac{P_1 V_1}{T_1} = \frac{P_2 V_2}{T_2}$$

where the subscripts distinguish between two sets of conditions. It applies to gases which behave nearly ideally and assumes that the number of particles of gas remains constant.

A still more general gas law takes into account the variations in the number of molecules of the gas. Again the law applies only to nearly ideal gases. The equation is $PV = nRT$ where n represents the number of moles of a gas and R is a constant whose value depends upon the units of P, V, and T.

INVESTIGATION

Purpose: To determine the molar volume of oxygen; to determine the percent potassium chlorate in a mixture.

Potassium chlorate can easily be decomposed in the presence of a catalyst, manganese dioxide, to form oxygen and potassium chloride $2KClO_3 \rightarrow 2KCl + 3O_2$. The oxygen is collected over water and its volume is measured at laboratory conditions. The weight of oxygen may be determined by the loss of weight in the potassium chlorate. By knowing the weight of oxygen and its volume at STP, the molar volume of oxygen can be calculated. It is also possible to determine the percent of potassium chlorate present in a mixture of substances, if none of the "contaminants" is volatile or reacts with either products or reactants.

Equipment:

- 1 Florence flask (500 ml)
- 1 ring stand
- 1 utility clamp
- 1 test tube
- 3 straight glass tubes (about 8 cm long)
- 1 beaker (600 ml)
- 1 glass tube, long enough to reach into the Florence flask, with a right angle bend at one end
- 2 stoppers: one-hole and two-hole
- 1 pinch clamp

Chemicals: potassium chlorate, $KClO_3$, solid; or mixture of $KClO_3$-KCl
manganese dioxide, MnO_2, solid

Procedure:

Set up the apparatus illustrated in Figure 12.

WARNING: Do not weigh any of these chemicals on paper.
Do not use glycerine to lubricate stoppers.

Weigh everything precisely.

Weigh everything at room temperature.

Place your 15 cm test tube in a 150-ml beaker; weigh and record their combined masses. Put either 1 gram potassium chlorate or the assigned weight of the unknown mixture into the test tube. Record the weight. Add about 0.1 gram manganese dioxide. Again record the weight. Assemble the apparatus as shown in the diagram. Note that the glass tube in the Florence flask which connects with the test tube is the short tube. This short tube protrudes only slightly below the rubber stopper in the flask.

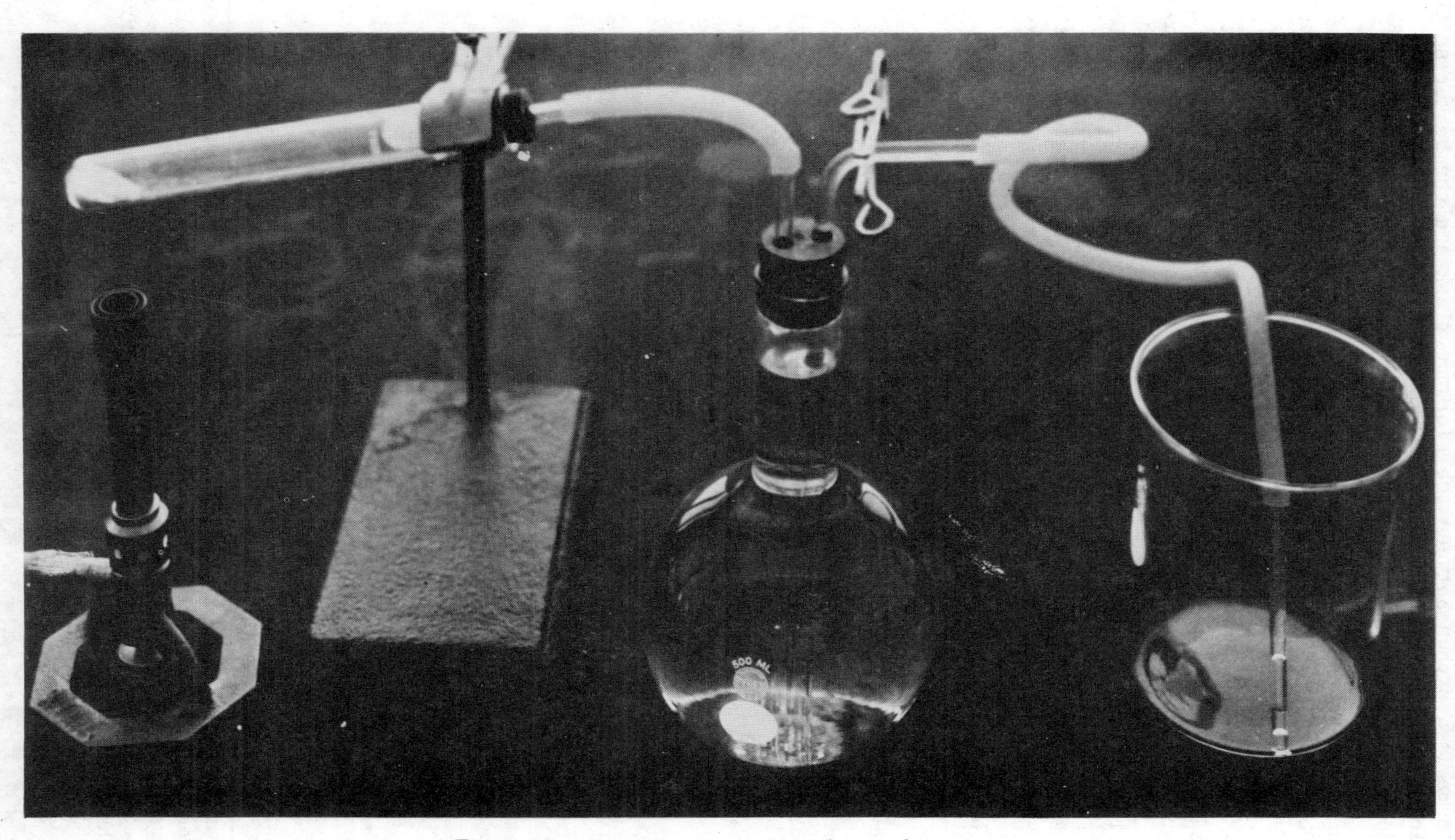

Figure 12. Apparatus for molar volume of a gas.

There is a syphon between the Florence flask and the beaker so that water can flow easily between them. If the assembled apparatus is set up correctly, it will be air tight, and no water will flow from the syphon.

Heat the potassium chlorate gently, holding the burner in your hand. The oxygen formed forces water through the syphon. When the reaction is complete the colume of water in the beaker remains constant.

*Do not remove the discharge tube from the beaker and break the syphon until the test tube has cooled to room temperature.** While the gas cools, the volume contracts and the syphon reverses until the system is at room temperature.

Clamp the rubber tubing leading to the beaker and remove the discharge tube from the beaker. Measure and record the volume of water in the beaker using a graduated cylinder. Measure and record the temperature of the oxygen by inserting a thermometer in the Florence flask above the water level.

Figure 13. Equalization of water levels in a Florence flask and a beaker.

* If, after the system has cooled, there is a large difference in water levels between the Florence flask and beaker, there may be a significant difference in pressure between the atmosphere and the oxygen in the flask. The pressures are equalized by raising or lowering the beaker until the water surfaces are at the same level and waiting 10 seconds. (See Fig. 13)

NAME______________________________________SECTION____________DATE____________

MOLAR VOLUME OF A GAS

DATA

1. a. barometric pressure ________
   b. temperature of the oxygen ________
   c. water vapor pressure at the measured temperature (table: Handbook of Chemistry & Physics) ________
   d. volume of oxygen collected at laboratory temperature and pressure ________

   Before reaction:

   e. weight of beaker, test tube, $KClO_3$-KCl, MnO_2 ________
   f. weight of beaker, test tube, $KClO_3$-KCl ________
   g. weight of beaker, test tube ________
   h. weight of $KClO_3$-KCl sample ________

   After reaction:

   i. weight of beaker, test tube, MnO_2, residue ________
   j. the weight of oxygen released during reaction ________

2. Using the proper number of significant figures, calculate the molar volume of oxygen showing all work clearly as follows.

   a. the moles of oxygen gas formed (use mass of gas from j.) ________

   b. the volume of oxygen at STP using the combined gas law ________

   c. the molar volume of oxygen calculated from experimental data (2b/2a) ________

   d. accepted value for the molar volume of a gas at STP ________

e. the percent error ________

3. The percent potassium chlorate in the unknown can be calculated if the number of moles of oxygen generated in the reaction is known. In this procedure there are two ways to calculate the moles of oxygen: 1) by using the weight of oxygen (see 'j' and 'a') and 2) by using the volume of oxygen generated at STP (see 'b') and the accepted value for molar volume. Using both values for the number of moles of oxygen generated, answer the following questions, showing all calculations and the proper number of significant figures.

	computations from weight	computations from volume
a. the number of moles of oxygen formed	______	______
b. the number of moles of potassium chlorate which decomposed to form this amount of oxygen	______	______
c. the number of grams of potassium chlorate present in the sample	______	______
d. the percent potassium chlorate in the sample	______	______

THOUGHT

1. Why must the apparatus cool to room temperature before the syphon between the Florence flask and beaker is disconnected?

2. In this experiment, it is possible to calculate the moles of oxygen from either the volume formed at STP or from the weight of oxygen generated. Which gives the more accurate value? Why?

NAME______________________________ SECTION__________ DATE__________

3. Explain the effect that failure to subtract the water vapor pressure from the total pressure would have on your determination of molar volume.

4. Why must the combined gas law and not the general gas law be used to calculate the volume of oxygen at STP in 2b (Data) in preparation for the determination of the experimental value for molar volume?

NAME________________________________ SECTION__________ DATE__________

Experiment 8

CONVERSION OF DISCARDED ALUMINUM TO A USEFUL PRODUCT (ALUM)

PRELIMINARY QUESTIONS

1. Write the formula for the double salt of potassium aluminum sulfate, alum.

2. What is the solubility of this salt?

3. Write balanced equations for the following chemical reactions:

   a) Aluminum metal reacts with excess potassium hydroxide, KOH and water to form soluble potassium aluminate, $KAlO_2$, and hydrogen gas.

   b) To potassium aluminate, $KAlO_2$, sulfuric acid, H_2SO_4, and water are added until aluminum hydroxide, $Al(OH)_3$, precipitates.

   c) Aluminum hydroxide is dissolved in sulfuric acid.

   d) Potassium aluminate, water, and sulfuric acid react to form alum, potassium aluminum sulfate.

4. Calculate the following:

a) the number of moles in 1 gram of aluminum ________

b) the number of moles in 4 grams of potassium hydroxide ________

c) the number of moles in 20 ml of 9 M H_2SO_4 ________

d) If the quantities above of aluminum, potassium hydroxide, and sulfuric acid are used to prepare potassium aluminum sulfate, which reactant is the limiting agent and determines the maximum amount of alum which can form theoretically?

e) What is the theoretical yield of alum expressed in grams?

Experiment 8

CONVERSION OF DISCARDED ALUMINUM TO A USEFUL PRODUCT (ALUM)

IDEAS

Although aluminum is the most abundant metal on the earth, it was not identified as an element until the first half of the nineteenth century because of the great difficulties in isolating it from its naturally occurring compounds. Aluminum compounds known as alums had been used since ancient times in the dyeing process, and were described almost two thousand years ago by Pliny, the Elder (who died in 79 AD during the eruption of Mt. Vesuvius), and by the medieval alchemist Geber in a famous manuscript of the early fourteenth century.

However, it was not until 1825 that impure aluminum was isolated by the Danish scientist C. Oersted, known for the discovery of the magnetic effect of an electric current. The German chemist F. Wohler, continuing these investigations, was more successful in his attempts to isolate aluminum, and was able to describe its properties.

The first pure aluminum and the first commercial process were produced by Saint-Claire Deville in 1854, separating aluminum from its chloride by the action of sodium. The first inexpensive commercial process for extracting pure aluminum was developed simultaneously in 1886 by Charles Hall, an undergraduate at Oberlin College, and Paul Heroult, in France. It was accomplished by the electrolysis of bauxite ($Al_2O_3 \cdot H_2O$) in fused cryolite (Na_3AlF_6). The resulting increased availability of aluminum coupled with its excellent metallic properties led to many new uses, from pots and pans to the airplane industry. Aluminum is present on earth in finite quantities and so must be recycled as efficiently as possible.

Note the international character of the researchers discussed above, their dependence on new discoveries and techniques, the pressure of accumulating knowledge that led to simultaneous discoveries, the impact on a new industry -- and the remarkable contributions of a 21 year old student.

Metallic aluminum is a very active metal which combines readily with the oxygen of the atmosphere (rusting) to form a transparent film of oxide on its surface. The oxide layer is relatively inert because, unlike iron rust,* the film is continuous and non-porous and thus protects the pure metal underneath from further chemical attack. The surface presented by the oxide is also largely resistant to chemical attack. The oxide which has amphoteric properties reacts with either acids or bases but because of its surface it does so slowly. Hence aluminum with its oxide coating has found many uses in food packaging. For example, aluminum sheets are often used in the freezer or refrigerator to wrap food where any chemical reaction between the

*Iron rusts when in the presence of both air and moisture. Iron rust, a hydrated form of Fe_2O_3, does not stick to the surface of the iron but flakes off continuously exposing new metallic surface to the atmosphere.

aluminum and the food is further slowed by the cold temperatures. It is used as trays for bread and cakes both of which are dry and neutral. For beverages which are acid (sodas) the inner liner of the can may be protected by a thin plastic film, or the beverage itself may contain a reducing agent such as tin (II) chloride to prevent oxidation (corrosion) of the inner surface of the can. There are a myriad of uses in the construction industry; in mechanical systems such as engines; in laboratories for the mirrors of reflecting telescopes; as the oxides to form rubies and sapphires used in lasers. Although the most abundant metallic element in the earth's crust (8.1%) as mentioned earlier, it is always found in combined form.

Since the hydrated oxides of aluminum are amphoteric (i.e., react with either acid or base) the following conventions are used in writing their formulas. If the hydrated oxide is considered for its basic properties the formula is usually written $AL(OH)_3$ showing the OH^- group. On the other hand, if the acidic properties are of interest, then the formula can be written H_3AlO_3. Forms of the aluminate ion are either $Al(OH)_4^-$ or AlO_2^-. The choice depends on the degree of hydration of the original material.

Originally the designation "alum" applied only to the double salts of potassium and aluminum sulfate. Defined in modern terms, however, the potassium ion may be replaced by the ammonium ion or any ion of the Group I elements (except lithium) and the trivalent aluminum ion can be replaced by either chromium (III) or iron (III). All alums characteristically have 12 molecules of water associated with the crystal. The general formula is $M(I)M(III)(SO_4)_2 \cdot 12H_2O$. Therefore, for potassium alum the formula is $KAl(SO_4)_2 \cdot 12H_2O$.

INVESTIGATION

Purpose: To observe amphoteric properties of aluminum; to prepare a pure product by crystallization.

Equipment:

- 1-150 ml beaker
- 1-250 ml beaker
- filter paper (coarse)
- glass stirring rod
- pan for ice bath
- hot plate
- string or thread

Chemicals:

- 1 g aluminum foil, Al
- 4 g potassium hydroxide, KOH
- 9 M sulfuric acid, H_2SO_4
- ethanol

Procedure:

Use scrap aluminum (i.e., foil, tray, or can). If a can is used, it will have to be polished with fine steel wool to remove both the painted decoration on one surface and a possible plastic film on the other.

Using small pieces of the aluminum weigh precisely about one gram of aluminum and 4 g of potassium hydroxide (KOH) on the centigram balance into a 250 ml beaker. Add enough deionized water to make 50 ml total solution. Mark the solution level with a glass marking pencil if your beaker has no calibration lines.

$$2Al + 2KOH + 2H_2O \rightarrow 2KAlO_2 + 3H_2$$

The beaker should become quite warm because the reaction of potassium hydroxide with water is exothermic. Stir the mixture with a glass stirring rod to disperse the heat. You should observe bubbles of hydrogen forming and the aluminum darkening and crumbling. If your reaction subsides and there is still evidence of aluminum present, heat gently for a few minutes at *low* heat on a hot plate. The reaction should take about 25-30 minutes. When the reaction is complete, there may still be dark insoluble flakes floating in the solution. These are not aluminum. The reaction is complete when there is no further evolution of hydrogen gas.

If necessary, add deionized water to the reaction mixture to bring the volume up to 50 ml and then filter the hot solution into a clean 250 ml beaker. Wash the residue on the filter paper with two 5-ml increments of deionized water. Add 20 ml of 9 M sulfuric acid *slowly, carefully and with stirring.* This is an exothermic reaction.

$$2KAlO_2 + H_2SO_4 + 2H_2O \rightarrow 2Al(OH)_3\downarrow + K_2SO_4$$

As the sulfuric acid neutralizes the solution, a thick gelatinous precipitate of aluminum hydroxide forms, but this should dissolve in the warm solution when the reaction medium becomes acid. Test the solution with litmus paper. If the aluminum hydroxide seems to resist dissolving in the acid, warm it gently, with stirring, on the hot plate. Any undissolved residues remaining after this treatment, should be filtered out. The filtrate contains the alum.

$$KAlO_2 + 10H_2O + 2H_2SO_4 \rightarrow KAl(SO_4)_2 \cdot 12H_2O$$

To produce a large crystal suspend a string from a rod so that it dangles about one cm below the surface of the solution. Cover the beaker with a piece of filter paper in which a few holes have been made. Secure the paper with a rubber band and allow the solution to stand undisturbed until the next laboratory period.

Filter the crystals and solution through coarse filter paper. (If a second crop of crystals is desired, allow the solution to continue to evaporate slowly. The 'second crop' crystals are usually smaller in size and may also contain occluded impurities.)

Prepare a solution of 5 ml water and 5 ml ethanol.* Wash your crystals with this solution by pouring the solution over the crystals on the filter paper. Put the crystals on a piece of dry filter paper and pat them gently with a paper towel to dry them. Assume they are dry when they no longer wet the filter paper or towel.

Transfer the dried crystals to a clean dry weighed vial and weigh them.

* The technique of "mixed solvents" is used when the product to be washed is very soluble in the washing medium. In this case, alum is very soluble in water but the addition of ethanol reduces its solubility.

The percent yield may be calculated as follows:

$$\frac{\text{experimental weight of alum (g)}}{\text{theoretical weight of alum (g)}} \times 100$$

Label your vial as follows:

your name
date
name of the product
grams of product
percent yield of the product.

Save your crystals for the experiment on crystal structure.

NAME______________________ SECTION__________ DATE__________

CONVERSION OF DISCARDED ALUMINUM TO A USEFUL PRODUCT (ALUM)

DATA

. Weight of 150-ml beaker and aluminum __________ g
. Weight of 150-ml beaker __________ g
. Weight of aluminum __________ g

. Weight of 150-ml beaker, aluminum and potassium hydroxide __________ g
. Weight of 150-ml beaker and aluminum __________ g
. Weight of potassium hydroxide __________ g

. Weight of dried crystals of alum and vial __________ g
. Weight of vial __________ g
. Weight of alum __________ g

How many grams of alum can form from this reaction theoretically? ______ g
Show calculations.

What is the percent yield? ______

THOUGHT

1. When an aluminum can finally disintegrates--after about 100 years--what is the most likely chemical product?

2. a. Why is it possible to recycle aluminum as opposed to coal or oil?

   b. What advantage does recovery and reprocessing of scrap aluminum have over the conversion of aluminum ore (bauxite) to aluminum in terms of energy conservation?

3. Why are the crystals of alum washed with a solution containing half water and half ethanol?

NAME____SECTION____________DATE__________

Experiment 9
CRYSTAL STRUCTURE

PRELIMINARY QUESTIONS

1. Define:

   a) unit cell

   b) crystal lattice

   c) coordination number

2. Make a drawing of a unit cell of each of these following structures:

   a) simple cubic structure

   b) body-centered cubic structure

   c) face-centered cubic structure

Experiment 9
CRYSTAL STRUCTURE

IDEAS

A crystal is a regular three-dimensional pattern of atoms or molecules. Although the theory that matter was made of atoms was not justified quantitatively until the nineteenth century, it was accepted by many individuals for thousands of years. A late-sixteenth century reference to the arrangement of atoms in a geometric pattern in a solid is by Thomas Hariot (1560-1621), an English mathematician and astronomer and the tutor of Sir Walter Raleigh. He imaginatively and concretely described the arrangement of atoms as follows:*

" 9. The more solid bodies have Atoms touching on all Sydes.
10. Homogeneall bodies consist of Atoms of like figure, and quantitie.
11. The weight may increase by interposition of lesse Atoms in the vacuities betwine the greater.
12. By the differences of regular touches (in bodies more solid), we find that the lightest are such, where euery Atom is touched with six others about it, and greatest (if not intermingled) where twelve others do touch euery Atom."

As early as 1669 it was announced by Steno that the angles of any one crystalline substance are constant. Modern crystallography was born in the late eighteenth century, with the work, among others, of the French priest, Rene Hauy (1743-1822). Working with minerals, he is said to have dropped a piece of calcite which broke into pieces of a specific geometric form (rhombohedral). Breaking the fragments further resulted in smaller new fragments of the same shape. Hauy proposed that the crystal was formed by addition of a series of unit cells and that the angles of the particular crystal and the ratio of the sides were constant. He also hypothesized that a relationship exists between crystal structure and chemical composition.

The determination of atomic arrangements in crystals and interatomic distances was accomplished in the early part of the twentieth century, when von Laue thought of using crystals to study the nature of x-rays. As a result of these experiments, x-rays were clearly shown to be similar in nature to light waves; in addition, the structure of crystals was further elucidated. According to a story attributed to Dr. Peter Debye, he, von Laue, and others, during coffee house discussions before World War I, considered the question of whether x-rays were wave-like or particle-like in nature. The wave-like nature of x-rays was experimentally demonstrated

*As quoted in L. Pauling, *College Chemistry*, W. H. Freeman and Company, San Francisco, 1964, p. 33.

when these rays were passed through a crystal and produced the characteristic diffraction pattern of electromagnetic radiation.

Significant contributions were made by W. H. Bragg and his son, W. L. Bragg, who formulated a quantitative relationship on the interference effects of x-rays produced by crystals. They were able to determine that crystals of sodium chloride contain alternating sodium ions and chloride ions rather than the sodium chloride molecule; this work greatly influenced ideas on the dissociation of ions in solution.

There are seven crystal systems. These are defined by the ratio of the length of the sides (a,b,c) and the size of the angles between them. In the cubic system all sides are equal to each other and all the angles are right angles. Within the cubic system there are several ways in which atoms can be arranged. These are the simple cubic, body centered cubic and face centered cubic arrangements, (cubic closest packing).

The smallest arrangement of atoms, ions, or molecules which, if endlessly repeated, would reproduce the crystal lattice is called the unit cell. To facilitate visualization of the structures at the atomic level, the atoms or ions are regarded as hard impenetrable spheres. They are arranged so that nearest neighbors touch each other (i.e., are bonded together). The total number of nearest neighbors is the coordination number or ligancy of that atom or ion. To simplify still further, in this experiment species which occupy the sites of the unit cell are either single atoms or ions. In nature these sites may be occupied as well by complex ions or molecules.

INVESTIGATION

Purpose: To build models of the three types of lattice structures associate with the cubic system; to compare the cubic closest packed and the hexagonal closest packed systems.

Equipment:
42 styrofoam spheres, 30 of one color and 12 of another
tray for spheres
one package of pipe cleaners, cut into 1 cm lengths
magnifying glass

Chemicals: Crystals of several different substances: salt, NaCl; potassium hexacyanoferrate (III), $K_3Fe(CN)_6$; alum, $FeNH_4(SO_4)_2 \cdot 12H$ and laboratory preparations

Procedure:

In this experiment atoms or ions will be represented by styrofoam spheres and the "bonds" between the atoms by the pipe cleaners. Build the structures exactly as described or pictured; do not add additional "bonds" which you may feel is necessary to provide greater stability or rigidity to the structure.

Two or more spheres may be joined together by inserting one end of the one cm length of pipe cleaner into one sphere, and then pushing a second sphere onto the wire until both spheres are in contact. When the spheres are to be detached, remove the wire carefully by pulling it straight out. Try to make as small a hole as possible.

I. Simple Cubic Structure:

Construct a group of simple cubic unit cells of the same color by building regularly around one simple cube, so that a large cube of 27 spheres is formed. Each side of this cube has nine spheres.

Use the pipe cleaner pieces to bond the spheres to each other only in planes which are perpendicular to each other.

Remove one sphere from the position in the center of your model, and insert another sphere of the same size but different color.

i. How many unit cells share this same sphere?

ii. What fraction of this sphere lies in each unit cell?

iii. How many nearest neighbors (ligancy, coordination number) does each sphere have?

iv. How many spheres make up one unit cell?

v. How many unit cells have you constructed with all 27 spheres? Express the value for the side (or edge) of the unit cell in terms of the radii of the spheres.

II. Body-Centered Cubic Structure:

Join 3 spheres together, using a different color for the center

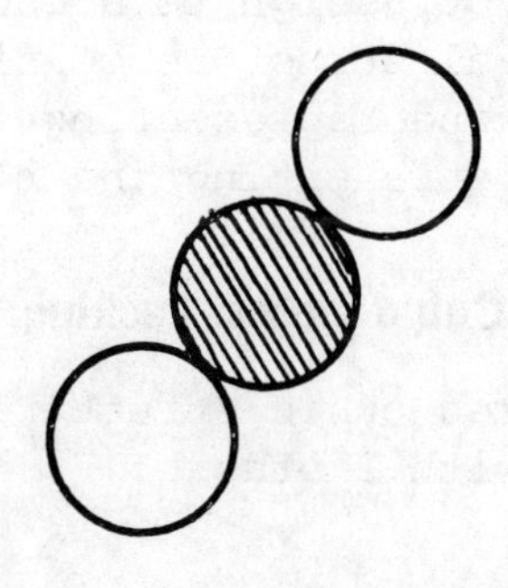

Visualize this line of 3 spheres as the body diagonal of a body-centered cube.

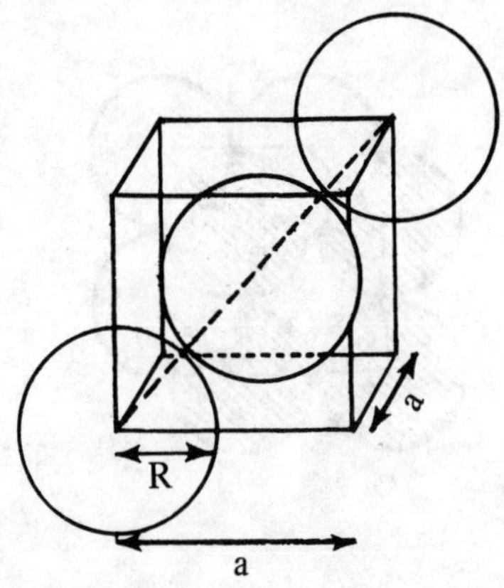

Maintain the same color for the corners, join the 6 spheres representing the *remaining* six corners of the cube to the center sphere. This sphere is also the center of the body-centered cubic structure. Now that a pattern has been established, continue to add more spheres in the color of the *center* sphere according to the same regular pattern. Visualize one corner of the cube as the center sphere of the next interpenetrating body-centered cube. Sixteen spheres are used for this construction. Since there is no bond between the corners of this cube, do not connect them with pipe cleaners.

i. In terms of the geometry of the cube, where do you find the spheres which touch each other?

ii. How many nearest neighbors (coordination number or ligancy) does each sphere have?

iii. What is the body diagonal expressed as radii of the spheres?

iv. What is the length of the side of this unit cell expressed as radii of the spheres?

v. Considering only *one* unit cell what is the ratio of the spheres in the corner to the spheres in the center?

vi. How many unit cells did you build?

vii. What formula would you assign to a chemical substance of a body-centered cubic crystal structure consisting of two different kinds of atoms or ions, one kind representing the corners of the unit cell (A) and the other kind representing the body centers of the cell (B)?

III. Face-Centered Cubic Structure Cubic Closest Packing:

A. To build the face-centered cubic structure, first join 3 spheres so that each one is in contact with 2 others.

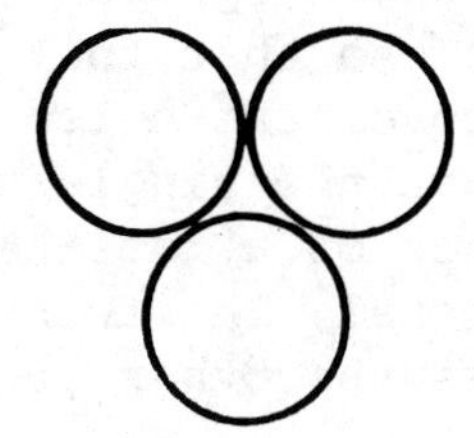

(Use spheres of the same color for this entire construction.) Place another sphere **on** the center hole of this equilateral base and attach it to one of the spheres. What is the name of this regular pyramidal construction? This shape is significant in terms of covalent bonding of the carbon atom.

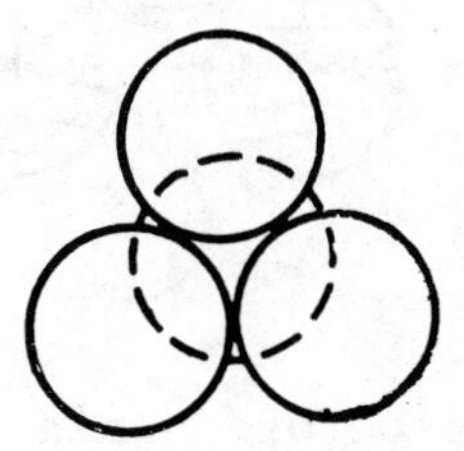

On the plane which you consider the base of the pyramid, add one additional sphere to each **space** available between the 3 base spheres.

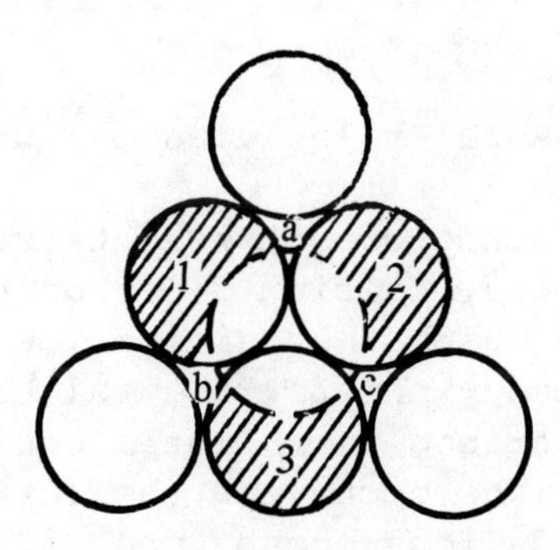

You now have an equilateral triangle of 3 spheres per side with one sphere perched on top in the center opening.

Make another identical construction of 7 spheres, using the same color as before. You are going to use these two identical sets of spheres to build a face centered unit structure.

First, turn one set over placing it on the table so that the single sphere becomes the support for the triangle of spheres. The angle between the table and the triangle of spheres is about 45°.

Next, holding the second set of spheres by the single sphere, fit them into the base resting on the table. Spheres 1, 2, 3 of the unit on the table form an equilateral triangle. The apex of the triangle in the drawing is pointed downward

(sphere #3). The second set of seven spheres is fitted into the first set to form a cube by holding the second set so that the apex of its equilateral triangle points upward and the spheres cover the holes a,b,c. When you examine the resulting structure, you will note that the single sphere of the second set which you are holding has become the corner of a face-centered cubic structure; it is also one end of the cell's body diagonal. The twin sphere of the first set of spheres which is resting on the table is the other end of the diagonal.

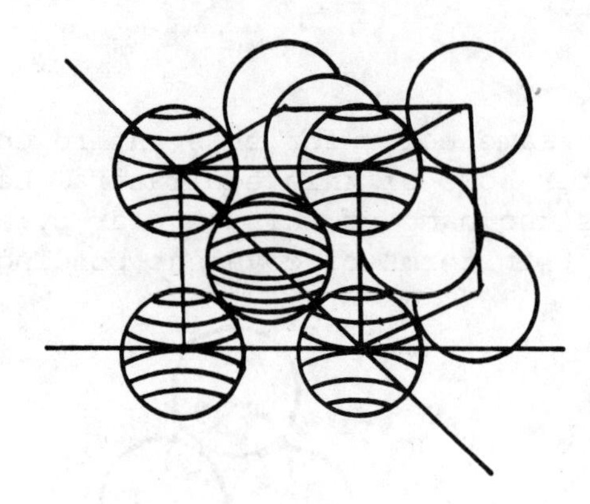

B. Build a face-centered structure again, but use one color for the 4 tetrahedrally arranged spheres in *one* of the two sets, with spheres of another color for the rest of the construction. Grasp these 2 sets (making up this face-centered structure) firmly. Can you see the "tetrahedral holes" (surrounded by the four spheres of different color) of the face-centered cube?

   i. How many tetrahedral holes does one unit cell have?

In certain crystal structures these "tetrahedral holes" are occuped by atoms or ions. For example, calcium fluoride can be considered as a face-centered cubic system in which the calcium ions occupy the corners and faces, and the fluoride ions occupy the tetrahedral holes. Can you also see how such a structure might be described as an interlocking face-centered simple cubic structure? The calcium ions occupying the lattice points of the face-centered structure and the fluoride ions occupy the lattice points of a simple cubic structure.

C. Construct a face-centered cube again, this time exchanging those spheres which represent the centers of the faces by spheres of *different color* and same size. This is best accomplished by using different-colored spheres for the 3 spheres which represent the base of the tetrahedral arrangement. Make 2 identical two-color sets and complete the face-centered cubic structure as before.

   i. What geometric form is described by the spheres occupying the places of the center of the faces only?

Construct this geometric double pyramid separately with other spheres. Insert 1 sphere of a different color (which may also be of different size) into the opening in the center of this pyramidal structure. This is an octahedral hole.

ii. How many octahedral holes are in the unit cell of this face-centered cube?

iii. What is the coordination number of this center sphere?

IV. Hexagonal Structure Closest Packing:

Build one set of 7 spheres of the same color, half a unit of the face-centered structure. Using the single top sphere surround this with 6 spheres of another color. You now have a hexagon with a sphere of different color in the center.

Turn this structure upside down so that the hexagon is the base and the equilateral triangle rests on it. Now place on top of the equilateral triangle 1 sphere; roll this sphere lightly to fit into the various available depressions (holes). You will notice that you can either put 1 sphere into the center of the equilateral triangle (position 1) or into the depression made by 3 spheres in the corners (position 2) (the only **possibility).**

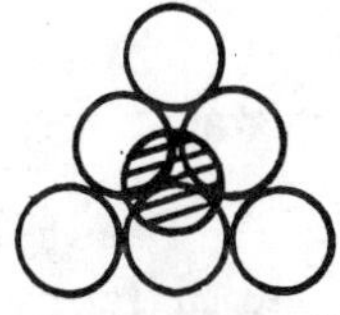

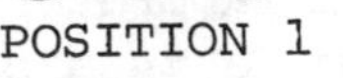

POSITION 1

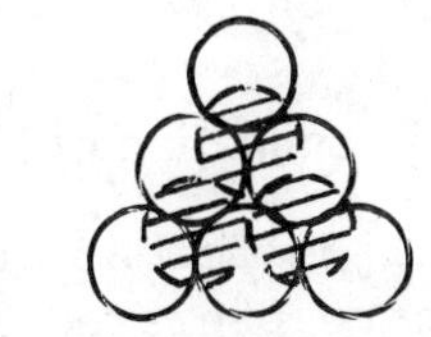

Possible placements for POSITION 2

Join 3 spheres together and place them first with one sphere in the center hole (position 1). Now observe that in this closest packing, the first layer (the hexagonal) and the third layer make up repeating patterns, with the spheres in this third layer directly on top of the first layer position. Now make another hexagonal layer of 7 spheres (1 in the center and 6 around it), and place this layer as a third layer (instead of the 3 spheres) exactly symmetrically on top of the first hexagonal layer. The triangular layer is sandwiched between the hexagonal layers. Lift up this whole arrangement, grasping it tightly by the middle spheres of the two hexagonal layers and hold this structure up to the light. You will see three spots where the light can penetrate freely through the openings in the arrangement. Obviously these three openings are located in the same place in each of the three layers. Place the structure on the table and look at it sideways. You will notice that the third layer is repeated. This sequence of layers, representing the hexagonal closest packing, can be expressed as the ABAB sequence, A and B representing different layers. Place the 3 spheres as the fourth layer and examine the structure again from the side, identifying the ABAB sequence.

V. Comparison of Closest Packed Systems

If, however, you use position 2, and place 3 spheres around the center hole, you will find that a different structure is produced. Build on these 3 spheres one half of a face-centered structure, as before, using these 3 spheres as the base of the tetrahedral arrangement. When completed, set the 3 spheres representing the base of the original tetrahedron (the equilateral triangle) in position 2. The lone sphere is in the fourth layer. See whether you can recognize the face-centered cubic structure on top of the first layer of different color. Remove the single top sphere and lift up these three layers, holding them against the light and looking through the arrangement. This time no light can penetrate through the structure, since all the holes are covered with spheres. This face-centered-cubic closest packed arrangement repeats itself only in the fourth layer and it is often referred to as the ABC ABC sequence of layers. To visualize this better, attach 6 spheres of the same color as the 6 spheres of the first layer around the 1 sphere which you removed before, which was the single top sphere in the center. This one is in the same position as the center of the base hexagon (first layer) the two centers of the hexagons being part of the diagonally opposite corners of the unit cell of the face-centered cubic structure. Can you see the ABCA layers? Examine your set viewing it from the side as well as looking down from the top.

Review the method of building the two closest packing structures in this fashion until you recognize the layers easily.

i. How many nearest neighbors exist in the hexagonal structure?

ii. How many in the face-centered-cubic closest packing?

Any crystal that you can see consists of billions of unit cells.

VI. Macroscopic Examination of Natural Crystals (Optional)

Use several different crystals at least one mm in size in each dimension which you may have prepared in the laboratory (e.g., alum) or which have been provided by your instructor. Examine them carefully with a magnifying glass or low power microscope. Report the following:

i. Name of the crystalline material.

ii. Can you tell if the crystal belongs to a cubic system?

iii. Are the angles between the planes 90°?

iv. If not, can you guess the approximate angle (45°, 30°)?

v. Reported crystal structure (Handbook of Chemistry and Physics)

NAME____ SECTION____________ DATE__________

CRYSTAL STRUCTURE

DATA

I. Simple Cubic Structure

i. How many unit cells share this center sphere? ________

ii. What fraction of this sphere lies in each unit cell? ________

iii. How many nearest neighbors (ligancy, coordination number) does each sphere have? ________

iv. How many spheres make up one unit cell? ________

v. How many unit cells have you constructed with all 27 spheres? Express the value for the side (or edge) of the unit cell in terms of the radii of the spheres. ________

II. Body-Centered Cubic Structure

i. How many nearest neighbors (coordination number or ligancy) does each sphere have? ________

ii. What is the length of the side of this unit cell expressed as radii of the spheres? ________

iii. Only certain spheres in this structure touch each other. How would you describe their position to the geometry of the cube? ________

iv. What is the ratio of the spheres in the corner to the spheres in the center considering only *one* unit cell? ________

v. How many unit cells did you build? ________

vi. What formula would you assign to a chemical substance of a body-centered cubic crystal structure consisting of two different kinds of atoms or ions, one kind representing the corners of the unit cell and the other kind representing the body centers of the cell? ________

III. Face-Centered Cubic Structure

i. How many octahedral holes are in the unit cell of the face-centered cube? (Hint: count whole holes and fractional holes.) ________

ii. How many tetrahedral holes does one unit cell have? ________

iii. What is the coordination number of the center sphere? ________

V. Hexagonal and Face-Centered Cubic Structures

i. How many nearest neighbors exist in the hexagonal structure? ________

ii. How many in the face-centered cubic closest packing? ________

VI. Optional: Results of the Examination of Natural Crystals

Name of crystalline material	I Alum	II	III
Angles between planes			
Suggested crystal system			
Reported crystal structure			

THOUGHT

1. Express the volume of a unit cell in terms of the radii of the spheres for each of the three cubic structures.

You may be able to apply what you have learned about crystals in this experiment to the solution of the following more sophisticated problems.

2. Calculate the fraction of the unit cell which is empty space for each of these three structures.

3.a. What kind of chemical formula would a compound have which consists of a face-centered crystal structure with different kinds of atoms in the corners (A) than in the center of the faces(B)?

b. What is the coordination number, closest nearest neighbors, of a same kind for (A)?

4. Could you suggest a structure for a pure substance with a chemical formula AB_2?

5. Chromium crystallizes in a body-centered cubic structure. The side of the unit cell is 0.288 nm. The atomic weight of chromium is 52.0 amu. Calculate the density in grams per cubic centerimeter of chromium.

6. Sodium chloride has a density of 2.17 g/cm^3 and a formula weight of 58.4 amu. The cubic unit cell is face-centered cube for Na^+ and for Cl^- ions contains four ion pairs (Na^+Cl^-). It has an edge equal to 5.63×10^{-8} cm. How many ion pairs are there in 58.4 g of sodium chloride?

7. Neon crystallizes in the cubic closest packing; the side of the unit cell is 0.452 nm. What is the atomic radius?

8. Cesium chloride has a structure which can be regarded as body-centered cubic; the unit structure has a side of 0.411 nm.

   a. What is the smallest Cs-Cl distance?

   b. What is the smallest distance between Cs-Cs and Cl-Cl atoms?

NAME____________________________________ SECTION____________ DATE____________

Experiment 10

IONIZATION AND CONDUCTIVITY

PRELIMINARY QUESTIONS

1. Define an ion.

2. Write the following formulas in molecular form and in ionic form:

   a) sodium chloride

   b) ammonium phosphate

   c) potassium nitrate

3. How can you determine experimentally whether or not a substance is present in ionic or molecular form when in aqueous solution?

4. If a substance is insoluble in water, what convention is used for writing the formula in an ionic equation? Explain.

5. What is the difference between a total ionic equation and a net ionic equation?

6. What is the difference between a strong acid and a weak acid? A strong base and a weak base?

7. What is the difference between the substances hydrogen chloride and hydrochloric acid?

8. Why is hydrogen chloride in an aqueous solution written as $H^+ + Cl^-$ and acetic acid as $HC_2H_3O_2$?

Experiment 10
IONIZATION AND CONDUCTIVITY

IDEAS

"Ion" or "wanderer" in Greek was one of several new words first used by M. Faraday (1791-1867) to describe new phenomena he had observed relating chemistry and electricity. Using the battery, invented in 1800 to pass a current of electricity through solutions containing dissolved compounds, Faraday found in 1832 that:

1. Certain substances he called electrolytes allowed electricity to pass through easily.

2. Free elements were liberated from their compounds by this process.

3. The mass of the element produced was related to the amount of electricity used, the atomic weight of the element, and its valence. This process of electrolysis had many implications.

It suggested that: 1) atoms in some compounds called electrolytes were charged, since they allowed current to pass through, while nonelectrolytes prevented the passage of current; 2) some atoms were negatively charged because they were attracted to the positive electrode where the charge was lost and the element was freed, while the converse was true for positively charged atoms (charged atoms or groups of atoms are called ions); 3) since a particular quantity of electricity, 96,500 coulombs, broke down one mole of any compound made up of atoms of valence one, then each single atom might be associated with a specific quantity or a "particle" of electricity.

The idea of the electron was born here. The explanation of chemical combination due to electrical charges on the atom was also proposed by Faraday, who wrote:

> It is impossible perhaps, to speak on this point without committing oneself beyond what present facts will sustain; and yet it is equally impossible, and perhaps would be impolitic, not to reason upon the subject. Although we know nothing of what an atom is, yet we cannot resist forming some idea of a small particle, which represents it to the mind; and though we are in equal, if not greater, ignorance of electricity . . . yet there is an immensity of facts which justify us in believing that the atoms of matter are in some way endowed or associated

with electrical powers, to which they owe their most striking qualities, and amongst them their mutual chemical affinity.*

These ideas were among the first to relate the subjects of chemistry and physics.

These inferences based upon Faraday's electrolysis experiments were no more than imaginative guesses until additional experiments were performed. Raoult's investigations on the freezing point depression of solutions led to predictions that agreed with observations only when nonelectrolytes were used. For electrolytes of the same concentration, the results were puzzling. For instance, the freezing point lowering produced by sodium chloride was just twice that predicted while the lowering produced by barium chloride or sodium sulfate was three times the predicted value. The Swedish chemist S. Arrhenius (1859-1927) concluded that the best explanation for this apparent anomaly was the breaking up or dissociation of electrolytes into ions in solution, for if one mole of an electrolyte contains two moles of ions, then the freezing point lowering should be twice that predicted. This suggestion was not accepted among chemists at that time, for the separation of such a stable substance as table salt into component parts and the existence of charged atoms seemed like foolish ideas. According to Dalton's theory, atoms were indivisible, and without structure.

It is now well established that compounds may exist in molecular or ionic form. As an example, carbon tetrachloride is a liquid in which four chlorine atoms are covalently bonded to a single carbon atom and the CCl_4 molecule is an independent and stable species. Sodium chloride, on the other hand, does not exist as individual NaCl molecules, but rather as independent Na^+ and Cl^- ions arranged in the crystal lattice so that the aggregate has electrical stability and the stoichiometry of one Na^+ per one Cl^- is maintained.

When dissolved in water, ionic compounds such as NaCl, KNO_3, and $CuSO_4$ break up into their constituent ions which are free to move about the solution as charged particles. Depending on their nature, molecular compounds may or may not break up into ions when they dissolve in water. Sugar in aqueous solution is molecular in form. It is called a nonelectrolyte. But HCl gas ionizes almost completely when dissolved in water:

$$HCl \xrightarrow{H_2O} H^+(aq) + Cl^-(aq)$$

The solvent medium is also a factor in determining whether the dissolved substance will ionize. In benzene, HCl gas will not dissociate into ions but will remain molecular in this organic solvent.

Substances which are almost completely dissociated in solution are said to be strong electrolytes. Substances which are mostly in molecular form in solution are said to be weak electrolytes. In these solutions only a small fraction of the molecules dissociate into ions. An example of a weak electrolyte is acetic acid, $HC_2H_3O_2$.

Chemical equations tell us what takes place during a reaction. The equations should not only indicate the reactants and the products but should also reflect the form in which each species is found. Equations should indicate whether the species involved are ionic or molecular, soluble or insoluble in water. We focus our attention on water as the solvent medium because usually it is aqueous solutions which are encountered in the laboratory.

*M. Faraday, *Experimental Researches in Electricity*, London, 1839-55; quoted in L. Taylor, *Physics, the Pioneer Science*, vol. II, Dover Publications, N. Y. 1959, pp. 627-628.

Writing Equations: Rules and Conventions

1. Molecular formulas are written for:

   a. non-electrolytes such as elements, gases, and most organic compounds. e.g., F_2, Cu, SO_2, HCl (when gaseous), CCl_4, C_2H_5OH (ethyl alcohol)

   b. weak electrolytes in aqueous solution: H_2O, $HC_2H_3O_2$ (acetic acid), HCN, NH_3, complex ions such as $Ag(NH_3)_2^+$

   c. Insoluble salts, e.g., AgCl, $BaSO_4$, $CaCO_3$

2. Ionic formulas are written for strong electrolytes in aqueous solution. e.g., hydrochloric acid is written as $H^+(aq) + Cl^-(aq)$

   sodium sulfate is written as $Na^+(aq) + SO_4^{2-}(aq)$

   potassium hydroxide is written as $K^+(aq) + OH^-(aq)$

3. If an acid or base is almost 100% ionized, the acid or base in solution is written as ions with a single arrow in the direction of the dissociation.

   e.g., (nitric acid) $HNO_3 \xrightarrow{H_2O} H^+(aq) + NO_3^-(aq)$

   Double arrows are used to indicate an equilibrium exists in solution. The arrows are of unequal length: the long arrow points to the species which predominates.

   e.g., (hydrocyanic acid) $HCN \underset{\longleftarrow}{\overset{H_2O}{\rightharpoonup}} H^+(aq) + CN^-(aq)$

From now on, it will be understood that in aqueous solutions all ions are hydrated. For simplicity we will omit writing aq (aqueous) after each ion or H_2O over the arrows.

Writing Equations: Reactions in Aqueous Solutions

Reactions involving ions in aqueous solution take place only if ions are removed from the field of action. This may be done by one or a combination of the following methods.

1. Formation of a precipitate. The extent to which a reaction goes depends on the solubility of the substance formed. If one or more of the products is insoluble, then the formation of these insoluble compounds drives the reaction toward completion.

   Total ionic equation:

   $$Ag^+ + NO_3^- + Na^+ + Cl^- \longrightarrow AgCl + Na^+ + NO_3^-$$

   Since the NO_3^- and Na^+ are unaffected in this reaction, they are referred to as spectator ions. It is preferable to omit them in writing the equation. The *net ionic equation* shows only those species taking part in the reaction.

Net ionic equation:

$$Ag^+ + Cl^- \rightarrow AgCl$$

In either case, the equation must show the proper form of the chemical substances, and must be balanced with respect to mass and charge.

2. Formation of a weakly ionized substance
e.g., a) water, b) a weak acid, c) a weak base, d) a weakly dissociated complex ion

a) Formation of water

$$H^+ + Cl^- + Na^+ + OH^- \rightarrow Na^+ + Cl^- + H_2O$$

Net ionic equation

$$H^+ + OH^- \rightarrow H_2O$$

b) Formation of a weak acid

$$Na^+ + C_2H_3O_2^- + H^+ + Cl^- \rightarrow HC_2H_3O_2 + Na^+ + Cl^-$$

Net ionic equation

$$C_2H_3O_2^- + H^+ \rightarrow HC_2H_3O_2$$

c) Formation of a weak base

$$Na^+ + OH^- + NH_4^+ + Cl^- \rightarrow NH_3 + H_2O + Cl^- + Na^+$$

Net ionic equation

$$NH_4^+ + OH^- \rightarrow NH_3 + H_2O$$

d) Formation of a complex ion

$$Ag^+ + NO_3^- + 2NH_3 \rightarrow Ag(NH_3)_2^+ + NO_3^-$$

Net ionic equation

$$Ag^+ + 2NH_3 \rightarrow Ag(NH_3)_2^+$$

3. Formation of a slightly soluble gas, (H_2S) or an unstable product which yields a gas on decomposing.
(e.g., H_2CO_3, H_2SO_3, HNO_2).

$$2Na^+ + CO_3^{2-} + 2H^+ + 2Cl^- \rightarrow 2Na^+ + 2Cl^- + H_2CO_3$$

then

$$H_2CO_3 \rightarrow H_2O + CO_2$$

Net ionic equation

$$CO_3^{2-} + 2H^+ \rightarrow H_2CO_3 \rightarrow H_2O + CO_2$$

4. Oxidation or reduction of an ion to a different oxidation state (this will be studied later in a separate experiment)

How will you know which substances are strong electrolytes or weak electrolytes? which are soluble salts and which insoluble salts? Use the summarizing information given below.

Summarizing Table for Acids and Bases

Strong Electrolytes	Weak Electrolytes
Acids:	
HCl, dilute H_2SO_4, HNO_3, HBr, HF	most other common acids
Bases:	
Hydroxides of Group I (NaOH, KOH, etc.)	most other common bases
Hydroxides of Group II except $Mg(OH)_2$	

Almost all salts are strong electrolytes. However, because insoluble salts produce few ions in the solution, they may conduct electricity poorly or not at all.

A few soluble salts are weak electrolytes. These dissolve in water as molecules, not as ions. A few examples of these are $CdBr_2$, $HgCl_2$, $Pb(C_2H_3O_2)_2$, and $CdCl_2$.

Summary of Solubilities

1. Common salts of K^+, Na^+, NH_4^+ are *all* soluble.
2. *All* nitrates, chlorates, and acetates of metal ions are soluble, except that $AgC_2H_3O_2$ is not.
3. Cl^-, Br^-, I^- salts of metals are soluble except Ag^+, Pb^{+2}, Hg_2^{2+}, Cu^+, and HgI_2.
4. All sulfates are soluble except Ba^{+2}, Sr^{+2}, Pb^{+2}, while Ca^{+2} and Ag^+ are moderately soluble.
5. Sulfides of Groups I, II in the periodic table and NH_4^+ are soluble. All others are insoluble.
6. Carbonates and phosphates of metal ions are insoluble except for Na^+, K^+, NH_4^+.

INVESTIGATION

Purpose: To determine the electrical conductivity of a variety of compounds; to learn to write net ionic equations; to learn to predict the extent of chemical reaction based on conductivity information and solubility tables.

Equipment:
2 30-ml beakers
1 hydrogen sulfide generator
electrical conductivity apparatus with light bulb

Chemicals: ethanol, C_2H_5OH; glycerine, $C_3H_5(OH)_3$; cyclohexane, C_6H_{12}; glacial acetic, $HC_2H_3O_2$; sodium chloride, NaCl, solid; potassium acetate, $KC_2H_3O_2$, solid; sugar, $C_{12}H_{22}O_{11}$; 0.1 M hydrochloric acid, HCl; 0.1 M acetic acid, $HC_2H_3O_2$; 0.1 M sodium hydroxide, NaOH; 0.1 M ammonium hydroxide, $NH_3 \cdot H_2O$; 0.1 M sodium chloride, NaCl; 0.1 M potassium nitrate, KNO_3; 0.1 M sulfuric acid, H_2SO_4; 0.1 M barium hydroxide, $Ba(OH)_2$; 0.1 M copper [II] sulfate, $CuSO_4$; hydrogen sulfide, H_2S, gas; hydrogen chloride gas in cyclohexane

Procedure:

A comparison between degree of ionization can be made by observing the electrical conductivity of pure substances and solutions. To perform the test, place about 10 ml of liquid or an equivalent volume of solid in a crucible or small beaker. Insert the wire electrodes into the substance. Observe the brightness of the light bulb. With good conductors the light will be bright, with poor conductors it will be feeble, and with nonconductors it will remain dark. Record the conductivity as good, poor, or none. Because the distance between the electrodes as well as the depth to which the electrodes are immersed affect the conductivity, *these tests should be made under uniform conditions*. Therefore, before you start to test be sure the electrodes are about 1 cm apart, and during the test immerse them to a depth of 1 cm. After each test, rinse the wires with distilled water and dry them with a paper towel. *DISCONNECT THE APPARATUS WHEN HANDLING THE ELECTRODES.*

Record all observations on the data pages.

I. Pure substances. Test conductivity of each of the following:

A. water

B. ethanol (C_2H_5OH)

C. glycerine [$C_3H_5(OH)_3$]

D. cyclohexane (C_6H_{12})

E. glacial acetic acid ($HC_2H_3O_2$)

F. solid sodium chloride

G. solid potassium acetate

H. molten potassium acetate

I. sugar ($C_{12}H_{22}O_{11}$)

II. Add enough deionized water to about 2 ml of the liquid or about 1 g of the solid substances listed in table II to make a total of 10 ml of the mixture. To prepare the hydrogen sulfide solution, bubble hydrogen sulfide gas for about 15 seconds into 10 ml of deionized water. This should produce a saturated solution, approximately 0.1 M. Test the conductivity of these solutions.

III. Compare the conductivity of these solutions:

A. HCl gas in water

B. HCl gas in cyclohexane

IV. Comparison of conductivity of acids and bases.

A. Compare the conductivity of an aqueous solution of 0.1 M $HC_2H_3O_2$ with an aqueous solution of 0.1 M HCl.

B. Compare the conductivity of an aqueous solution of 0.1 M NaOH with an aqueous solution of 0.1 M NH_3.

V. Ionic reactions. Determine the conductivity of each of the following solutions individually. Then carefully mix equivalent amounts of each pair and test for conductivity of the mixture. Record the conductivities in the table.

The reaction indicated in (e) below requires special care. Immerse the electrodes in the barium hydroxide solution. Add the sulfuric acid dropwise and stir the solution well after each addition. Observe the changes in conductivity of the solution as the sulfuric acid is added.

a. 0.1 M NaCl and 0.1 M KNO_3

b. 0.1 M HCl and 0.1 M NaOH

c. 0.1 M $HC_2H_3O_2$ and 0.1 M NH_4OH

d. 0.1 M $CuSO_4$ and H_2S

e. 0.1 M H_2SO_4 and 0.1 M $Ba(OH)_2$

NAME________________________ SECTION____________ DATE__________

IONIZATION AND CONDUCTIVITY

DATA

I. Write the formula for each in molecular form when there is no conductivity and in ionic form if there is conductivity. Use Table I below.

Table I

Substance	Conductivity	Formula
A. water		
B. ethanol		
C. glycerine		
D. cyclohexane		
E. glacial acetic acid		
F. solid sodium chloride		
G. solid potassium acetate		
H. molten potassium acetate		
I. sugar		

II. Characterize the conductivity of each substance below as strong, weak, or non-conducting.

Table II

	Conductivity
A. ethanol	
B. glycerine	
C. acetic acid	
D. sodium chloride	
E. potassium acetate	
F. sugar	
G. hydrogen sulfide	

i. What kind of compounds are glycerine, and sugar as determined by the conductivity tests? ______________

ii. What kind of compounds are sodium chloride and potassium acetate?____

iii. Compare the conductivities of each type of compound.__________

III. i. Compare the conductivity of HCl gas dissolved in water and HCl gas dissolved in cyclohexane.

ii. To what do you ascribe the difference in results between the two solutions of HCl?________________________________

iii. Illustrate the results with an ionic equation.

IV. A. i. Conductivity of 0.1 M HCl ____________

ii. Conductivity of 0.1 M $HC_2H_3O_2$____________

Write equations for the dissociation of HCl and $HC_2H_3O_2$ in water, using double arrows of unequal length when only partial ionization is involved.

iii. ____

iv. ____

B. i. Conductivity of 0.1 M NaOH____________

ii. Conductivity of 0.1 M NH_3____________

Write equations for the dissociation of NaOH and NH_3 in water, using double arrows when only partial ionization is involved.

iii. ____

iv. ____

v. Which is the stronger acid, HCl or $HC_2H_3O_2$?_________

vi. Which is the stronger base, NaOH or NH_3?__________

vii. What is the appropriate way to write the formula in aqueous solution of NaOH?

____

of $HC_2H_3O_2$?______________________________

NAME________________________SECTION__________DATE__________

V. Record conductivities. Use the following notation; 'C' for conducting solutions; 'N' for nonconducting solutions; and 'P' for solutions which conduct poorly.

Table V

	Reagent A	Reagent B	Reagent A	Reagent B	Mixture
a.	0.1 M NaCl	and 0.1 M KNO_3			
b.	0.1 M HCl	and 0.1 M NaOH			
c.	0.1 M $HC_2H_3O_2$	and 0.1 M $NH_3 \cdot H_2O$			
d.	0.1 M $CuSO_4$	and 0.1 M H_2S			
e.	0.1 M H_2SO_4	and 0.1 M $Ba(OH)_2$			

i. Give net ionic equations for any reactions which occur on mixing reagents A and B.

ii. On the basis of your net ionic equations, give a brief but clear explanation for (c) and (e) in terms of the observed conductivity.

THOUGHT

Write net ionic equations for the following chemical reactions taking place in aqueous solution; use the summary tables p. 117 to determine which substances are insoluble and write these in molecular form.

a. calcium carbonate and nitric acid

b. silver nitrate and calcium chloride

c. copper [II] acetate and hydrogen sulfide

d. zinc [II] sulfate and hydrogen sulfide

2. In each of the above reactions predict the electrical conductivity of the products and the reactants.

NAME____ SECTION____________ DATE__________

Experiment 11

OLECULAR WEIGHT DETERMINATION BY FREEZING POINT DEPRESSION

PRELIMINARY QUESTIONS

1. Define molality.

2. What is meant by freezing point depression?

3. Calculate the freezing point of two aqueous solutions, one containing 10% calcium chloride by weight, the other 10% methyl alcohol by weight. Disregard interionic attraction.

4. If 0.512 g of a substance is dissolved in 7.03 g of naphthalene, the freezing point is lowered from 80.6 to 75.2°C. What is the molecular weight of the substance? The molal freezing point lowering for naphthalene is 6.9°C/m.

5. An unusual skyscraper in Pittsburgh is supported by 18 hollow exterior columns. The columns are filled with water to prevent buckling in case of fire. If fire heats a column, it is cooled by convection currents circulating to the warm area. Potassium carbonate added to the water will prevent freezing during cold weather. The columns require about 500,000 gallons of 38% potassium carbonate solution, or 750 tons of potassium carbonate. Calculate the freezing point of this solution in Celsius and Fahrenheit degrees.

1. [illegible]

3. [illegible]

[illegible]

[illegible]

Experiment 11

)LECULAR WEIGHT DETERMINATION BY FREEZING POINT DEPRESSION

IDEAS

It is a commonplace observation that salt water freezes at a lower temperature than fresh water. In the eighteenth century this phenomenon was studied by C. Blagden (H. Cavendish's assistant), who noted that for solutions of the same components, the freezing point depression was proportional to the concentration. Investigating this phenomenon many years later, F. M. Raoult (1830-1901) in 1881 found that "the quantities of different substances which depress the freezing point (of any one solvent) by an equal amount are those which the chemist calls molecular quantities."* Raoult's rule permitted the molecular weight of many substances to be calculated. If one mole of any dissolved solute depresses the freezing point of one kg of a particular solvent by a constant value, then observations of the freezing point depression produced by an unknown solute permit the determination of its molecular weight. Similarly, an elevation of the boiling point was also observed. Raoult found that his generalization applied only to nonelectrolytes. The puzzling behavior of electrolytes led to the hypothesis of S. Arrehenius that these compounds dissociate into ions in solution. For instance, one mole of calcium chloride dissociates into 3 moles of ions.

Raoult also investigated the vapor pressure of solutions and found in 1886-1887 that "one molecule of a fixed, non-saline substance, in dissolving in 100 molecules of any volatile liquid, diminishes the vapor tension of the liquid by a nearly constant fraction of its value."** Known as Raoult's law, this generalization is described in modern texts in slightly different terminology: the vapor-pressure lowering of a solvent is proportional to the mole fraction of the solute particles. Vapor pressure depression causes both elevation of the boiling point and lowering of the freezing point.

As mentioned above, the molecular weight of an unknown solute can be calculated if we know (a) its weight, (b) the weight of the solvent, (c) the experimental freezing point depression of the solution, and (d) the molal freezing point depression constant of the solvent. The molal freezing point depression constant of a solvent is the decrease in freezing point of a solution containing one mole of a solute per kg of solvent. For water this constant is –1.86°C, for naphthalene it is -6.9°C, and for benzene it is –5.08°C. Since we know that one mole of a substance dissolved in one kg of naphthalene will lower its freezing point 6.9°C, we can determine the

* Quoted in F. J. Moore, *A History of Chemistry*, 3rd ed., McGraw-Hill Book Company, Inc., New York, 1939, p. 263.

** Quoted in J. R. Partington, *A Short History of Chemistry*, McGraw-Hill Book Company, Inc., New York, 1939, p. 333.

molecular weight of a substance dissolved in it if we measure the freezing point depression due to a known amount of solute.

This can be expressed according to the equation $t = K_f m$, where K_f is the molal freezing point depression constant, m is molality, and t is the difference between the freezing point of the pure solvent and the freezing point of the solution, an experimentally determined value.

The freezing points of a pure solvent and a solution may be determined by the use of a cooling curve. The data for the cooling curve are the temperatures of a cooling liquid collected at regular time intervals. In general these curves are constructed by plotting temperature on the y-axis and time on the x-axis.

INVESTIGATION

Purpose: To determine the molecular weight of an unknown solute using freezing point depression data.

Equipment:

- 1 400-ml beaker
- 1 15-cm Pyrex test tube
- 1 cork (to fit test tube)
- 1 looped wire stirrer to be made from heavy-gauge wire, preferably copper
- 1 thermometer, 110° C
- 1 ring stand
- 1 wire gauze
- 1 ring
- 1 clamp
- 1 burner
- 1 150-ml beaker

Chemicals:

Solvent: naphthalene ($C_{10}H_8$) molal freezing point depression 6.9°C.

Solute: various organic compounds as unknowns (e.g., anthracene, or other similar substances which dissolv in naphthalene).

Procedure:

See note for waste disposal of naphthalene and naphthalene solutions at en of procedure.

Set up the apparatus. Make sure that the side of the cork is grooved with a triangular file to fit the looped wire stirrer, so that the stirrer can be moved freely up and down. The thermometer bulb should be inserted to within 1.5 cm of the bottom of the test tube. The thermometer scale should be visible above 70°C.

1. All equipment must be clean and dry. To support the test tube during the weighing use a 150 ml beaker and record the weight of this assembly. Record all weights precisely.

   Weigh precisely about 10 g of naphthalene into a clean, dry test tube. Assemble the apparatus and put enough water into the 400 ml beaker so that the water level is above the naphthalene in the test tube. Heat the water and naphthalene to well above the melting point of the solid (85° to 90°C). Remove the burner and allow the naphthalene to cool slowly, stirring constantly until the freezing point has remained constant for about 3 minutes

Keep the test tube containing the naphthalene in the water while cooling. Record temperature readings each minute during cooling. When the naphthalene freezes, the thermometer is firmly embedded, so to remove the thermometer, the naphthalene must be softened by gentle heating. A certain amount of naphthalene will adhere to the thermometer. Be careful not to lose any of this naphthalene. Proceed to the second part of the experiment after two satisfactory melting points have been determined.

2. Into the test tube containing the naphthalene weigh precisely about 1 g unknown. The naphthalene-unknown mixture is heated *carefully* in the water bath until the solution is perfectly clear. (Naphthalene vapors are extremely flammable.)

   The apparatus is reassembled and the water bath is heated to well above the melting point of the solution. Remove the burner and allow the solution to cool slowly with *constant* stirring. Keep the test tube containing the solution in the water while cooling. Again record temperature readings each minute during cooling.

Waste disposal for naphthalene and naphthalene solutions must be in special containers. Do NOT dump into the sink.

NAME______________________________ SECTION__________ DATE__________

MOLECULAR WEIGHT DETERMINATION BY FREEZING POINT DEPRESSION

DATA

1. a. weight of 150 ml beaker, test tube, and naphthalene _________ g

   b. weight of 150 ml beaker and test tube _________ g

   c. weight of naphthalene _________ g

   Record data for cooling curve of pure naphthalene in Table I below.

Table 1

Cooling of Pure Naphthalene
(Record temperature each minute)

Temperature °C trial 1	Time trial 1	Temperature °C trial 2	Time trial 2

2. a. weight of beaker, test tube, naphthalene and unknown _________ g

   b. weight of beaker, test tube, and naphthalene (line 1a above) _________ g

   c. weight of unknown _________ g

Record data for cooling curve of mixture in Table 2 below.

Table 2

Cooling of Mixture of Naphthalene and Unknown Solute

Temperature °C trial 1	Time trial 1	Temperature °C trial 2	Time trial 2*

Plot the temperature-cooling curves for pure naphthalene and the naphthalene-unknown solution on the same graph.

3. a. Freezing point, pure naphthalene, trial 1 __________ °C

   b. Freezing point, pure naphthalene, trial 2 __________ °C

   c. Freezing point, mixture, trial 1 __________ °C

   d. Freezing point, mixture, trial 2* __________ °C

   e. Freezing point depression, Δt __________ °C

   Δt is the difference between the freezing point of the pure solvent and the freezing point of the solution. You will find that the cooling curves of the solution will have several "plateaus" or breaks in the line. This is due to the increase in the concentration of the solute as the solvent crystallizes, which in turn causes further depression of the freezing point. Hence, the freezing point of the solution for molecular weight determination is the temperature at which the *first* break in the curve is detected.

4. Calculations:

   a. Weight naphthalene __________ g

   b. Weight unknown solute __________ g

* If time permits, two trials should be completed. In that case, consult your instructor.

NAME__________________________________SECTION____________DATE_______________

c. Freezing point depression, Δt (3e.) __________ °C

d. Molal freezing point depression, $K_f = \frac{6.9°C}{\text{mole solute/kg solvent}}$ for naphthalene

e. Calculated molality ($m = \Delta t/K_f$) unknown solute______________
Show calculations.

f. Molality can also be expressed as follows:

$$m = \frac{(\text{g solute})}{(\text{g naphthalene})} \times \frac{(1000 \text{ g naphthalene})}{(\text{kg naphthalene})} \times \frac{(\text{mole})}{(\text{M.W. unknown, g})}$$

From this the molecular weight can be calculated by rearranging terms.

The calculated molecular weight for the unknown is ____________
Show calculations clearly.

THOUGHT

1. After the solute has dissolved in the solvent, and the solution is clear, will loss of small amounts of this solution effect the experimental results? Explain.

2. What effect will the following errors have on your *molecular weight* determinations? Note the effect and explain the reasons.

   a. If the solvent is lost through evaporation or carelessness.

   b. If the weight of solute is less than the recorded weight because you lost some in transfer after weighing.

c. If your thermometer reads about 1°C lower than it should along its entire range.

d. If part of the solute fails to dissolve.

3. a. Could you use benzene, or water as solvents for molecular weight determinations?

b. What are the freezing point depressions for these substances?

benzene ______________

water ______________

NAME_______________________________________SECTION___________DATE__________

Experiment 12
TITRATION: ACID-BASE

PRELIMINARY QUESTIONS

1. Define the following terms.

   a. titration

   b. indicator

   c. molarity

   d. normality

   e. equivalent weight for acid-base reactions

2. What weight of HCl gas is dissolved in one liter of 0.1 M hydrochloric acid?

3. a) How many moles of HCl gas are contained in 25 ml of a 0.1 M solution of the acid?

   b) How many moles of NaOH are required for the neutralization of the solution in (a)?

   c) If the concentration of the NaOH is 0.2 M, how many ml of NaOH are required for this neutralization?

4. a) How many moles of protons (H^+) are in 25 ml of 0.1 M H_2SO_4?

b) How many ml of 0.2 M NaOH are needed for complete neutralization in (a)?

5. a) What is the gram-equivalent weight of H_2SO_4 in a reaction in which all the hydrogens are neutralized?

b) How many equivalents are contained in 10 grams of H_2SO_4?

6. a) What is the gram-equivalent weight of NaOH?

b) What is the gram-equivalent weight of $Al(OH)_3$?

At the end point in an acid-base reaction the number of equivalents of acid are equal to the number of equivalents base.

7. Does the preceding statement hold for mole relationships? Explain your answer using the two examples given in questions 3 and 4.

NAME____________________ SECTION____________ DATE____________

8. a) What is the normality of a solution that contains 3.65 grams of HCl gas in one liter?

b) How many equivalents of HCl are contained in 25 ml of a 0.1 N solution of the acid?

9. If 25 ml of a 0.25 N NaOH solution were titrated with 30 ml of an HCl solution,

a) How many equivalents of NaOH were used?

b) How many equivalents of HCl were present?

c) What is the normality of this HCl solution?

10. What are the gram equivalent weights of each of the following:

a) H_3PO_4: in titration with NaOH to yield NaH_2PO_4.

b) H_3PO_4: in titration with NaOH to yield Na_2HPO_4.

c) $Al(OH)_3$: in titration with sulfuric acid to yield $Al_2(SO_4)_3$.

11. Write a balanced net ionic equation for the complete neutralization of oxalic acid by sodium hydroxide.

12. How many equivalents for neutralization are in each of the following solutions?

a) 25 ml of 0.2 N H_2SO_4

b) 25 ml of 0.2 M H_2SO_4

c) 25 ml of 0.2 M $Al(OH)_3$

13. Calculate the normality of the base:

a) 25 ml of NaOH is used to neutralize 20 ml of 0.10 N H_2SO_4.

b) 25 ml of $Ca(OH)_2$ is used to neutralize 20 ml of 0.10 N HCl.

Experiment 12
TITRATION: ACID-BASE

IDEAS

"Acids" are mentioned as early as the thirteenth century in an Indian manuscript and in a Spanish treatise on alchemy. Hundreds of years later, the Belgian physician J. Van Helmont (1579-1644) used the terms acid and alkali and wrote of producing water by neutralizing acids with chalk. Although Van Helmont believed in alchemy, he called himself a chemist; he is said to represent the transition from alchemy to chemistry.*

In the seventeenth century, two physician-chemists, Silvius and Tachenius, theorized that the chemical action of the living body depended upon the reaction between acids and alkalis. Tachenius also wrote that "all salts are composed of an acid and an alkali."** As it was noted that effervescence is an indication of the union of an acid and an alkali, each was recognized by the effervescence produced in the other. Dissatisfied with these vague definitions, M. Bertrand wrote in 1683 that "an acid is a liquid body composed of small firm and pointed particles, slightly resembling very fine and delicate needles,"*** which explained the prickling sensation of acids on the tongue and effervescence. "Alkali, on the contrary, should be a solid earthy body the particles of which have between their junctions pores of different structure," enabling their penetration by acids.

Although R. Boyle did not agree with these definitions, he could not formulate a better one himself.

However, because Boyle had greater interest and skill in the laboratory, he developed a systematic series of indicators that showed by color change if substances were acidic, basic, or neutral. It has been said that he especially liked the use of color changes to indicate that a chemical reaction had occurred because of their effectiveness in showing the presence of new substances and because they spectacularly impressed the casual visitor. His tests became widely known. For instance, he noted that syrup of violets and all other blue vegetable substances were turned red by acids.

When investigating the quantitative combination of acids and bases, H. Cavendish found in 1766 that the same weights of a particular acid are neutralized by different weights of various bases. These combining weights of the base he termed equivalent weights. The German chemist, J. Richter, in the last decade of the eighteenth century, continued and improved upon this work,

*See IDEAS from the experiment on the Analysis of a Copper-Nickel Alloy for a definition and discussion of alchemy.

**Otto Tachenius, *Hippocrates Chemicus,* 3rd ed., Lugd. Bat., 1967, p. 8; as quoted in J.M. Stillman, *The Story of Alchemy and Early Chemistry*, Dover Publications, New York, 1960, p. 391.

***Reflections nouvelles sur l'acide et sur l'alcali, par M. Bertrand, Lyon, 1683, ibid., p. 401.

strongly convinced that chemistry is part of applied mathematics. He coined the word "stoichiometry" and defined it as "the art of measuring the chemical elements." Although chemists had earlier attempted quantitative measurements, Richter was the first to generalize the information which had accumulated over the preceding one hundred years.

Richter's tables of the quantities of various bases or acids which would neutralize a given quantity of a specific acid or base were improved by others, and the resulting tables included a list of the proportions by weight in which acids and bases combine. Based upon careful and patient experimentation, these combining weights, later called equivalent weights, had no theoretical meaning at the time they were determined. Indeed, these tables were compiled before the atomic theory was given a firm foundation by Dalton. Not until more than a half century later were the apparently many separate strands of thought firmly woven into the fabric of chemistry. The beginning chemistry student learns and uses the theoretical meaning of the terms equivalent weight, atomic weight, and molecular weight, but should be aware of the genesis of these ideas from laboratory investigations.

Titrimetry is a very accurate method for determining the amount of substance present in solution. An "unknown" is analyzed by causing it to react completely with a measured amount of reagent of exactly known concentration.

In aqueous solution there are several types of reactions which can be utilized in titrimetric (volumetric) analysis. These are:

a) acid-base reactions (in which there is an exchange of protons)
b) oxidation-reduction reactions (in which electrons are exchanged)
c) precipitation reactions (in which an insoluble salt is formed)
d) complex-formation (in which a weakly ionized complex is formed)

Not every reaction, however, lends itself to this type of analysis. A successful titration requires that the reaction take place rapidly and go almost to completion (i.e., that the yield of products is almost 100%). It is also necessary to know when the stoichiometric amounts of reactants have been mixed; that is, when the end point is reached. Many methods are employed to signal that the reaction is complete. The most common one is the use of indicators whose color changes coincide with the end point.

In this acid-base titration you will determine the equivalent weight of an unknown acid. To a fixed weight of the acid which is dissolved in water, several drops of phenolphthalein indicator are added. The solution is colorless. Carefully, a measured volume of standardized sodium hydroxide solution is added from a buret and at the end point the color instantly changes to a light pink.

Analysis of any unknown solid or solution requires the use of a primary standard or a standardized solution. A primary standard is a very pure and stable reagent of definite composition. It may be accurately weighed directly into the titration flask; or an accurately weighed sample of the standard may be put into a volumetric flask and diluted to the mark with water forming a standard solution of known concentration. An accurately measured volume of this standard solution is added to the titration flask.

Sodium hydroxide is not a primary standard. Since it reacts with both water and carbon dioxide in the atmosphere, its composition is indefinite. However a standardized solution of the base can be prepared by dissolving pellets of sodium hydroxide in water and using approximate measurements. The exact concentration of the solution can be found by titration with a solution of oxalic acid which is the primary standard.

Concentration Units: Titrations are a type of volumetric analysis using carefully measured volumes of the solutions. It is therefore important to learn how to express quantities of reagent when these are dissolved in water. One precise way of referring to solution concentrations is the use of the term molarity.

A. Molarity

Molarity (M) is defined as the number of moles of solute (n) per liter of solution.

$$M = \frac{\text{number of moles of solute}}{\text{liters of solution}} = \frac{\text{moles}}{V}$$

If a solution is 0.1 M in HC1, it means that 0.1 mole of hydrogen chloride gas is dissolved in one liter of solution forming 0.1 M hydrochloric acid.

For any volume of solution the number of moles of dissolved solute in that volume may be calculated by the relationship:

$$\text{number of moles of solute} = \frac{\text{(moles)}}{\text{(liter)}}\ \text{(liter)} = M\ V \qquad \begin{aligned} M &= \text{molarity} \\ V &= \text{volume in liters} \end{aligned}$$

Sometimes a more concentrated solution must be diluted to a lower concentration. This means adding water, so that there are fewer solute particles in a given volume. Suppose you have to prepare 150 ml of a 0.2 M HNO_3 and you have only 6 M HNO_3. The calculations are straightforward if you recognize that

$$\text{number of moles before dilution} = \text{number of moles after dilution}$$

$$\text{number of moles} = \frac{\text{(moles)}}{\text{(liter)}}\ \text{(liter)} = VM$$

$$(V_{conc})\ (M_{conc}) = (V_{dil})\ (M_{dil})$$

$$(V)\ \frac{(6M)}{(\ell)} = (150\ \text{ml})\ \frac{(\ 1\ )}{(1000\ \text{ml})}\ \frac{(0.2\ \text{moles})}{(\ell)}$$

$$\begin{aligned} V &= 0.005\ \text{liter} \\ &= 5\ \text{ml} \end{aligned}$$

Five ml of 6M HNO_3 is diluted with enough water to make 150 ml total solution.

Acid-Base Titrations: In an acid-base titration the essential reaction is:

$$\underset{\text{Acid}}{HA} + \underset{\text{Base}}{BOH} \rightarrow H_2O + \underset{\text{Salt}}{BA}$$

When an acid and base have exactly neutralized each other, there is no excess acid or base left over.

Consider the reaction between a strong acid and a strong base:

$$HCl + NaOH \rightarrow H_2O + NaCl$$

Since hydrochloric acid and sodium hydroxide are almost 100% dissociated, the net ionic equation is

$$H^+ + OH^- \rightarrow H_2O$$

Now consider the reaction:

$$H_2SO_4 + 2NaOH \rightarrow 2H_2O + Na_2SO_4$$

and the net ionic equation:

$$2H^+ + 2OH^- \rightarrow 2H_2O$$
$$H^+ + OH^- \rightarrow H_2O$$

In both examples, the neutralization involves the formation of one mole of water from one mole of H^+ and one mole of OH^-. Therefore, it is often convenient to define another concentration unit based on moles of H^+ rather than moles of solute.

B. Normality

Normality is defined as the number of equivalents in a liter of solution. The concepts of normality and equivalents can be applied to several types of reactions. For this experiment, we will use the definitions appropriate to acid-base reactions.

One equivalent of an acid is that quantity of acid which furnishes one mole of H^+.

The number of equivalents is calculated from the expression:

$$\text{number of equivalents} = \frac{\text{weight of acid}}{\text{equivalent weight of acid}} = \frac{\text{grams of acid}}{\text{g equivalent weight}} = \frac{\text{g}}{\text{g eq wt}}$$

(Note the similarity between the definition of number of equivalents and the expression for the calculation of number of moles.)

The gram equivalent weight of an acid is defined as:

$$\text{gram equivalent weight} = \frac{\text{gram molecular weight}}{\text{number of } H^+ \text{per formula replaced in reaction}} = \frac{\frac{\text{g}}{\text{mole}}}{\frac{\text{eq}}{\text{mole}}} = \frac{\text{g}}{\text{eq}}$$

Similarly, one equivalent of base is that weight of base which will react with one mole of H^+.

$$\text{The gram equivalent weight of base} = \frac{\text{gram molecular weight}}{\text{number of } H^+ \text{per formula accepted}} = \frac{\frac{\text{g}}{\text{mole}}}{\frac{\text{eq}}{\text{mole}}} = \frac{\text{g}}{\text{eq}}$$

Normality is defined as the number of equivalents of acid or base per liter of solution.

$$N = \text{normality} = \frac{\text{number of equivalents of acid or base}}{\text{volume in liters}}$$

We may summarize the relationships in neutralization titrations that have been introduced as follows:

$$\text{number of eq acid} = \text{number of eq base}$$

$$\frac{\text{g acid}}{\text{g eq wt acid}} = N_A V_A = N_B V_B = \frac{\text{g base}}{\text{g eq wt base}}$$

Normality and molarity are simply related.

$$N = \frac{\text{number of eq}}{\text{liter of solution}} = \frac{(\text{number of moles})}{(\text{liter of solution})} \frac{(\text{number of eq})}{(\text{mole})} = M \frac{(\text{number of eq})}{(\text{mole})}$$

A 1 M HCl solution is also 1 N, but a 1 M solution of H_2SO_4 is 2 N since there are two equivalents per mole of sulfuric acid.

It should be pointed out that the value of normality and equivalents is intimately tied to the particular reaction involved. As an example, consider the neutralization of phosphoric acid.

$$H_3PO_4 + 3NaOH \rightarrow Na_3PO_4 + 3H_2O$$

$$H_3PO_4 + 3OH^- \rightarrow PO_4^{3-} + 3H_2O \qquad \text{(net ionic equations)}$$

In this reaction all the protons in the phosphoric acid have been neutralized and the normality is three times the molarity since there are three equivalents per mole of acid.

However, it is not necessary to remove all the protons completely.

$$H_3PO_4 + NaOH \rightarrow NaH_2PO_4 + H_2O$$

$$H_3PO_4 + OH^- \rightarrow H_2PO_4^- + H_2O \qquad \text{(net ionic equations)}$$

In this case, only one proton per H_3PO_4 molecule has been displaced and the normality and molarity of a phosphoric acid solution in this reactiion are identical.

INVESTIGATION

Purpose: To find the equivalent weight of an unknown acid by titration with a standardized base.

Attention

To obtain satisfactory results, you must follow the directions given below *with the utmost care*. The techniques of volumetric analysis require complete attention to the following details:

1. Precise weighing
2. Accurate recording of the measurements you make
3. Thorough cleanliness of your apparatus
4. Preventing the dilution of prepared solutions, by rinsing the receiving container (such as burets, pipets, and receiving flasks) first with water and then with small (5 ml) samples of the solution which is to be transferred to the container.
5. The addition of the solution delivered from your buret, without the loss of a single drop, into the Erlenmeyer flask containing the solution to be titrated.

Equipment: :

2 50-ml burets	1 250-ml Erlenmeyer flask
1 250-ml volumetric flask	1 wash bottle
1 Florence flask	1 50-ml beaker
1 300-ml storage bottle	3 125-ml Erlenmeyer flasks

Chemicals: oxalic acid, $H_2C_2O_4 \cdot 2H_2O$, solid
sodium hyrdoxide, NaOH, solid
phenolphthalein, 0.1% solution

Procedure:

I. Standard Acid Solution

Because the utmost care is exercised in the preparation of standard solutions, the normality is known to the maximum number of significant figures that can be justified by your measuring apparatus.

Calculate how much oxalic acid you will require to prepare 250 ml of a 0.5 N solution. The formula of chemically pure oxalic acid used in this experiment is $H_2C_2O_4 \cdot 2H_2O$. All the acid hydrogens in oxalic acid will be neutralized. **Remember to include the water of crystallization in your calculations of the molecular weight of the oxalic acid.**

After your instructor has approved your calculations use a previously **weighed beaker to weigh out precisely an amount which is close to your calculated value and record the weight.** *It is not necessary to spend time trying to weigh out your exact calculated value.*

Record all data on data pages.

Add a small amount of deionized water to the oxalic acid crystals you have weighed and transfer them with care into the volumetric flask. Use a stream of ionized water from your wash bottle to remove any trace crystals that may have adhered to the glass container, adding these washings to your volumetric flask. Add enough water so that your flask is about 1/2 full. Stopper the flask and agitate the contents until complete solution has been achieved. Oxalic acid is not very soluble, and it may take some time to dissolve completely.

Fill the flask to just below the mark. Bring the solution up to the mark by adding the last few drops with a medicine dropper or a wash bottle. Stopper your volumetric flask tightly so that no solution will be lost before complete mixing is achieved. Mix the contents of the flask by inverting it about 25 times, until no streaming lines can be observed. Your solution is now ready to be used as a standard. Since its concentration is known, you must be careful not to do anything to change it.

The normality of your solution can be calculated from the following:

$$\text{number of equivalents of oxalic acid} = \frac{\text{grams of oxalic acid}}{\text{g eq wt}}$$

$$N = \frac{\text{number of equivalents of oxalic acid}}{\text{liter of solution}}$$

Using the method outlined above, calculate the normality of your acid solution. Record this value in the table. Transfer this solution into a clean, dry storage bottle and stopper tightly. Label the bottle, as follows: contents, concentration, your initials, and the date of preparation.

II. Preparation of Base Solution

Calculate the number of grams of solid sodium hydroxide required to prepare 500 ml of 0.5 N sodium hydroxide. It is best to use a freshly opened bottle of this reagent. Because of its reactivity with the atmosphere, it is absolutely necessary to *close the reagent bottle tightly as soon as you have weighed your sample.*

a. USE ONLY GLASS CONTAINERS FOR SODIUM HYDROXIDE.

b. HANDLE THE PELLETS WITH A SCOOPULA. NEVER ALLOW THE PELLETS TO COME INTO CONTACT WITH YOUR SKIN.

c. NEVER, NEVER LEAVE ONE SINGLE SODIUM HYDROXIDE PELLET ON THE WEIGHING BENCH. IF YOU HAVE ACCIDENTALLY SPILLED ANY, REMOVE WITH A SCOOPULA IMMEDIATELY AND DISCARD IN THE SINK, WASHING DOWN WITH WATER. WIPE CLEAN THE AREA OF SPILLAGE, USING MOIST PAPER TOWELING WHICH IS TO BE DISCARDED.

Weigh a 150 ml beaker covered with a watch glass. Place an amount of sodium hydroxide close to your calculated quantity in the beaker and weigh again. Record your data on the data sheet.

Transfer your pellets to a 500-ml Florence flask and add 100 ml of deionized water. Because this is an exothermic reaction, the solution becomes very hot. Therefore, swirl the contents of the flask to prevent the pellets from sticking to the bottom, to assist solution, and to keep the heat of solution evenly dispersed. Once solution has been effected, add 400 ml of deionized water, close with a *rubber* stopper, and mix thoroughly. Label the bottle with the name of the contents, its approximate normality, the date, and your initials.*

III. Titration of the Base Solution

Using detergent and long brushes, clean two burets. When the buret is thoroughly clean, rinse three times with tap water and twice with deionized water. Rinse one buret twice with 5-ml portions of base solution, rejecting rinsings through the tip. Fill the buret including the tip with sodium hydroxide solution and clamp the buret vertically in an appropriate support.

Repeat the above procedure with the second buret, this time rinsing and filling with oxalic acid. Clamp this buret in place and prepare for titration. Inspect your burets to make sure there are no trapped air bubbles along its entire length, including the tip. Air bubbles can be removed by allowing a small amount of solution to drain through the tip into a small beaker used to hold waste solutions. The solutions in both burets should be at or slightly below the zero mark. Reading the bottom of the meniscus, record the level of the liquids in both burets to within 0.02 ml. Remove any drop adhering to the tip of the buret by touching it with a piece of filter paper.

* These solutions can be stored and the analysis completed in the following laboratory period.

You are now ready to titrate. From your buret, put about 20 ml (four signigificant figures) of oxalic acid into a clean but not necessarily dry Erlenmeyer flask. Place a piece of white paper under the flask. Add 2 drops of phenolphthalein indicator, and then add about 15 ml of base solution from your buret, swirling the flask. Titrate carefully to the end point, with a continuous swirling motion. When the pink color of the indicator lingers a second or two, you are near the end point. Now add the base dropwise, washing down the walls of the flask with a small amount of deionized water from your wash bottle as each drop is added. At the end point one drop of base will turn the solution a pink color which persists at least 30 seconds. If you have added too much base and exceeded the end point, add about one ml oxalic acid to the titration flask. Record this added volume exactly and enter the new final volume reading of oxalic acid in the table. The solution should now be colorless. Titrate to the end point again adding base in one drop increments.

Record the final volumes of both burets. Repeat this experiment twice more, using about 25 ml and 30 ml of oxalic acid, exactly measured.

IV. Equivalent Weight of an Unknown Acid

Obtain from your instructor an unknown solid acid. Your instructor may give you a weight range for your unknown which will permit a titration of 20 to 40 ml of your base. If, however, no information is given, a rough titration can be made on one gram of the unknown acid. Then the weight of acid can be calculated which will enable you to perform the titration properly.

Three samples of acid to be titrated are weighed accurately in each of three clean, dry and weighed 125-ml Erlenmeyer flasks. Approximately 25 ml of deionized water is added to each sample of solid acid to dissolve it. Add 3 drops of phenolphthalein and titrate each solution to a permanent pink end point. Record all data on the data page.

NAME____SECTION______________DATE____________

TITRATION: ACID-BASE

DATA

I. Preparation of Standard Oxalic Acid Solution

Calculation for the preparation of 250 ml of 0.5 N oxalic acid solution with $H_2C_2O_4 \cdot 2H_2O$ Use the proper number of significant figures in the recording of data and the calculation of results.

Data:

Weight of beaker and oxalic acid ____________

Weight of beaker ____________

Weight of oxalic acid ____________

Normality of oxalic acid ____________

Show method of calculation of your standard oxalic acid:

II. Preparation of Base Solution

Weight of beaker, watch glass and sodium hydroxide ____________

Weight of beaker and watch glass ____________

Weight of sodium hydroxide ____________

III. Titration of Base Solution

	Trials		
Buret readings, acid	I	II	III
Final reading			
Initial reading			
Volume used			

Normality of your standard acid

	Trials		
Buret readings, base	I	II	III
Final reading			
Initial reading			
Volume used			

To calculate the normality of your base:

$$N_{base} = \frac{V_A N_A}{V_B}$$

	Trials: I	II	III
Calculated normality of your base	______	______	______

Average normality of your base ____________

Show method of calculations:

IV. Equivalent Weight of an Unknown Acid

	Trials I	II	III
Weight of Erlenmeyer and unknown acid			
Weight of Erlenmeyer			
Weight of solid unknown acid			
Base buret reading final			
Base buret reading initial			
Volume of base used, ml			

Calculate the equivalent weight of the acid, using the following relationships:

$$(\text{volume of base, ml})\frac{(\text{number of equivalents of base})}{(1\text{ liter})}\frac{(1\text{ liter})}{(1000\text{ ml})} = \frac{\text{grams of aci}}{\text{equivalent wt of}}$$

	Trials: I	II	III
Calculated g equivalent weight	______	______	______

Average gram equivalent weight ____________

NAME______________________________ SECTION__________ DATE__________

THOUGHT

1. a. What is the effect on the normality of the standard oxalic acid solution, if some of the water evaporates?

   b. What effect will this have on the experimental value of the normality of the sodium hydroxide?

   c. What effect will this have on the experimental value of the equivalent weight of the unknown acid?

2. Do you have to know the exact volume of water used to dissolve the unknown solid acid for titration? Explain.

3. Why is the water of hydration included in the calculation for the molecular weight of oxalic acid?

NAME______________________________ SECTION__________ DATE__________

Experiment 13

CID-BASE TITRATION USING THE pH METER. DETERMINATION OF THE pKa OF A WEAK ACID.

PRELIMINARY QUESTIONS*

1. Define pH

2. a) Why is there a change in pH during the course of an acid-base titration?

   b) What is the pH at the endpoint of the titration of a strong base and a strong acid?

   c) Why is the pH at the endpoint of the titration of a strong base with a weak acid different?

3. a) What is a standard reference electrode?

   b) What is its function?

4. What does a glass electrode measure?

---

* Some of the information needed to answer these questions may be found in the experiment on Acid-Base Titration Using a Visual Indictor.

5 a) How much oxalic acid is needed to make up 250 ml of a 0.25 M solution?

b) Why can oxalic acid be used as a primary standard?

6. How much sodium hydroxide must be used to make up 500 ml of a 0.5 M solution?

7. How is it possible to obtain the pK_a of a weak acid from the pH of its solution?

Experiment 13

ACID-BASE TITRATION USING THE pH METER. DETERMINATION OF THE pKa OF A WEAK ACID.

IDEAS

The recognition of the electrical properties of matter and their eventual utilization in chemistry has a long and curious history. Accumulating knowledge on charging by friction aroused enormous public interest by the middle of the eighteenth century, and large crowds were attracted to "electrical" spectacles. In Paris, for example, a large audience watched as 700 monks joined hand to hand simultaneously leaped into the air as they took a shock from Leyden jars*. To experience the shock from a Leyden jar or a static electricity generator, though often hazardous, became an eagerly sought after experience.

Allowing charged objects to discharge represents a momentary surge of electrical charge. The production of a steady stream of charges, or an electrical current, stemmed from the investigations of the physiologist Luigi Galvani (1737-1798), who noticed the peculiar twitching of frogs' legs hanging from brass hooks that were attached to an iron grating. This was especially noticeable during thunderstorms or when a static electricity generator was in operation. The spasm was perceptible when the leg was simultaneously in contact with the brass hook and the iron fence. Galvani wrote, ". . . our hearts burned with desire to investigate also the force of electricity in quiet times during the day."** After many experiments, he concluded that "possibly the electricity was present in the animal itself." Galvani believed that this "animal electricity" was a "vital-force," a force associated with life.

Alessandro Volta (1745-1827) disproved this hypothesis in his famous paper "On the Electricity Excited by the Mere Contact of Conducting Substances of Different Kinds" published in 1800, in which he described the striking results he obtained from different arrangements of two different metals separated by a moist conductor. For instance, he alternated as many as 60 metal disks of two kinds, for example, zinc and copper, in a pile, separating each by paper that had been soaked in salt water. Connecting the first and last disk (which were of different metals) by means of a conductor produced the same kind of effects as the Leyden jar or the electrostatic generator, except that in Volta's setup the arcing or shock that was produced continued for a long time. "This endless circulation or perpetual motion of the electric fluid may seem paradoxical, and may prove inexplicable, but it is nevertheless real," Volta wrote. He also de-

* A glass jar that was coated on the inside and outside by a conducting material, such as metal foil. A large charge could be produced on the inside, an equal and opposite charge on the outside, and the contact of the inside and outside by means of a conductor produced an electrical discharge.

**L. Galvani, "The Electric Current," in W. F. Magie, ed., *A Source Book in Physics*, Harvard University Press, Cambridge, Mass., 1969, p. 423.

vised what he called "the crown of cups" to produce a current. It consisted of a row of several cups, each containing salt water, all joined together "in a sort of chain by means of metallic arcs," of which one arm might be copper and the other zinc. The different metal arms were immersed in the solutions, and touching the first zinc arm and simultaneously the last copper arm produced a shock. Experimental setups of this kind which produce electric current from chemical reactions are still called voltaic or galvanic cells or piles. Shortly after Volta's paper was published, William Nicolson (1753-1815) used the apparatus described by Volta to decompose water into hydrogen and oxygen. This represents one of the first uses of the voltaic cell to produce electrolysis, a chemical reaction.

When a new device is produced, it often stimulates a new series of investigations which result in the formulation of new definitions and concepts. The voltaic cell is an excellent example of this. For instance, André M. Ampère (1775-1836) in his 1822 paper "Experiments on the Electrodynamical Phenomena" tried to define and differentiate between what he called "electric tension" and "electric current." The former is now called potential difference, meaning difference in electrical potential energy per unit charge, for which the unit "volt" is commonly used. Interestingly, "tension" still survives in the term "high tension line." Ampère writes of the electric current as a continuous "electromotive action" produced in a closed curcuit. The unit of electric current, defined as the rate of flow of charges, or coulombs per second, was named the ampere in his honor. Several years later, Georg Simon Ohm (1789-1854) formulated the relationship known as Ohm's law: For a given conductor in an electrical circuit, the current flowing (I) is directly proportional to the potential difference ($\mathcal{E}$). The proportionality constant is the resistance (R) of the conductor.

Michael Faraday's (1791-1867) contributions to the field of electricity spanned 23 years that began with his first published paper in 1823, "Experimental Researches in Electricity." In his paper on electrolysis (1834), which contained some of the earliest ideas relating the fields of physics to chemistry, Faraday used new words he had coined to describe the phenomena he had observed. Some of these words were the following: ion (meaning "wanderer" in Greek), cation, anion, anode, cathode, electrode, and electrolyze. When passing a current of electricity through solutions containing dissolved compounds, Faraday found that 1) certain substances he called electrolytes allowed electricity to pass through easily; 2) free elements were liberated from their compounds by this process; 3) the mass of the element produced was related to the amount of electricity used, the atomic weight of the element, and its valence. This process of electrolysis had many implications.

These results suggested to Faraday that 1) atoms in some compounds, called electrolytes, were charged, since they allowed current to pass through, while nonelectrolytes prevented the passage of current; 2) some atoms were negatively charged because they were attracted to the positive electrode where the charge was lost and the element was freed, while the converse was true for positively charged atoms; 3) since a particular quantity of electricity, 96,500 coulombs, broke down one mole of any compound made up of atoms of valence 1, then each single atom might be associated with a specific quantity or a "particle" of electricity.

The idea of the electron was born here. The explanation of chemical combination due to electrical charges on the atom was also proposed by Faraday, who wrote:

> It is impossible perhaps, to speak on this point without committing oneself beyond what present facts will sustain; and yet it is equally impossible, and perhaps would be impolitic, not to reason upon the subject. Although we know nothing of what an atom is, yet we cannot resist forming some idea of a small particle, which represents it to the mind; and though we are in equal, if not greater, ignorance

> of electricity . . . yet there is an immensity of facts which justify us in believing that the atoms of matter are in some way endowed or associated with electrical powers, to which they owe their most striking qualities, and amongst them their mutual chemical affinity.*

In all these investigations on the chemical production of an electric current (the voltaic cell) and the use of the current to cause chemical changes (the electrolytic cell), the very fundamental question arose of how to account for the production of an electric current by two unlike metals in a conducting solution, or separated by a moist conductor. Hermann Walther Nernst (1864-1941), in a paper published in 1889, making use of the contributions of Jacobus Henricus van't Hoff, Svante Arrhenius, and Wilhelm Ostwald, formulated a theory on the production of current based on the solution pressure of the ions in contact with a metallic electrode. Comparing solution pressure to vapor pressure, Nernst wrote "we have, in evaporation and solution, processes which can be regarded as entirely analogous . . . . Metals tend to drive their ions into solution through forces analogous to the usual vapor pressure.**

The idea of "electrolytic solution pressure" was useful, wrote Nernst, even though the physical meaning was rather uncertain. From mathematical relations which he formulated relating potential difference to the solution pressure of the metal and its ions,*** Nernst predicted that all metals have the ability to go into solution to form traces of ions. Although the original form of Nernst's equation has been revised to the modern form**** used in this experiment, it is still known as the Nernst equation.

In this section three methods of analysis which depend upon specific electrical properties of solutions are discussed.

The potentiometric and pH titrations both depend upon the change in electromotive force (potential difference) between the two electrodes of the system when the concentration of species varies. Conductometric titration depends upon the amount of electric current passing through a solution of an electrolyte under fixed voltage as the concentration of the various species changes.

Potentiometric Titrations

Potential drop (electromotive force, emf) is measured in volts. A potential difference between two electrodes can be produced if one electrode is placed in an oxidizing agent and the other is placed in a reducing agent. The two solutions are separated by a porous barrier which is permeable to ions. Electrons travel through an external wire to complete the circuit. As the reaction proceeds, the changes in the concentrations of the reacting species produces a change in the voltage which may be used to study the course of the reaction.

A simple and direct way to measure changes in potential would be simply to immerse a pair of electrodes in the solution to be analyzed and measure the potential drop (voltage) between them with a voltmeter. However, this method is unsatisfactory because accurate voltage readings cannot be obtained. The voltmeter draws some current for its own operation. Second, a reaction occurs at the electrodes as a result of current flowing through the solution and changes

* M. Faraday, *Experimental Researches in Electricity*, London, 1839-1855; quoted in L. Taylor, *Physics, the Pioneer Science*, Vol. II, Dover Publications, New York, 1959, pp. 627-628.

**W. H. Nernst, "On the Process of Solution of Solid Bodies," in H. Leicester and H. Klickstein, eds., *A Source Book in Chemistry*, Harvard University Press, Cambridge, Mass., 1968, p. 495.

***$E = 0.860T \ 1n \ P/p \times 10^{-4}$ volt. E is the potential difference; P, the electrolytic solution pressure of the metal; p, the pressure of the ion; and T, the absolute temperature.

****See page 206. Ion activities or concentrations in the cell are used rather than solution pressure.

the species in the solution; last, the solution with its electrodes (the cell) also has an internal resistance which changes as current passes through the solution. Ideally it would be best to measure potential differences when there is no, or practically no, current passing through the solution. Ohm's law predicts that if the resistance (R) is very high for a given voltage ($\mathcal{E}$) then the current (I) will be low.

A device which measures voltages under these conditions is a potentiometer. One of the common commercially available instruments is the pH meter, which uses an amplified current and has scale readings calibrated in both millivolts and pH. By applying Ohm's law, $\mathcal{E} = IR$, and assuming an extremely small current of 10^{-7} ampere, the resistance of the unknown cell can be as high as 10 ohms with this instrument.

The electrodes to be used are of two kinds: a reference electrode and an indicator electrode. The reference electrode maintains a constant voltage during the titration. The potential of the indicator electrode changes as concentration changes. The difference between these two potentials can be read directly on the scale of the instrument.*

A common reference electrode is the saturated calomel electrode. It consists of a layer of mercury covered with solid mercurous chloride and potassium chloride. A saturated solution of potassium chloride rests on top of this solid layer. A platinum wire leads from the mercury layer; a salt bridge** allows the exchange of ions between the electrode, containing potassium chloride solution and saturated mercurous chloride, and the solution being analyzed.

$$\text{Half-reaction: } Hg_2Cl_2 + 2e \longrightarrow 2Hg^o + Cl$$

The potential of this electrode compared to a standard hydrogen electrode is approximately 0.242 volts. Temperature changes in the solutions have some effect upon the potential of this reference electrode.

The indicator electrode for redox systems is usually an inert metal electrode such as a platinum electrode. Since the voltage of the reference electrode is constant, the indicator electrode reflects changes in potential resulting from changes in the concentration of the species in the solution during a titration, as previously mentioned.

An example of a redox system is the oxidation of iron (II) ions to iron (III) ions by dichromate ions. Writing the half-reactions:

$$(Fe^{2+} \longrightarrow Fe^{3+} + e) \times 6 \qquad \text{Oxidation, anode}$$

$$Cr_2O_7^{2-} + 14H^+ + 6e \longrightarrow 2Cr^{3+} + 7H_2O \qquad \text{Reduction, cathode}$$

$$6Fe^{2+} + Cr_2O_7^{2-} + 14H^+ \longrightarrow 2Cr^{3+} + 7H_2O + 6Fe^{3+} \qquad \text{Sum of half-reactions}$$

The potential during the titration depends upon the concentration of all the species in the solution.*** In potentiometric titrations it is not necessary to know the absolute value of the cell potential. If the reference electrode maintains a constant potential and the indicator

* There are other available "combined" or single electrodes which are units containing both reference and glass electrode. It is used in exactly the same way as the separate electrodes. Each electrode has two leads instead of one.

** A salt bridge is a glass tube plugged on each end by porous membranes and filled with a saturated salt solution, commonly potassium chloride. One end is immersed in one cell half, the other end in the other cell half to form a bridge allowing the passage of ions between cell halves.

*** For a more complete discussion, see D. Skoog and D. West, *Fundamentals of Analytical Chemistry*, Holt, Rinehart Winston, Inc., New York, 1969, p. 395.

electrode is sensitive to the potential changes, all the plots of the potential changes during the titration under specific conditions will be similar in shape regardless of the voltage of the reference electrode.

Potentiometric titrations are feasible because the systems depend upon a spontaneous transfer of electrons between electrodes. It is possible, therefore, to relate this kind of reaction to the thermodynamic concept of free energy. The following discussion shows the relationship between free energy and electrode potentials, as shown by a form of the Nernst equation.

Derivation of the Nernst Equation

Free energy can be measured if a system does work on its surroundings or if the surroundings do work on the system, and the pressure and temperature of the system remain constant. The usual symbol for free energy is G,* the quantity being measured experimentally is the change in free energy, ΔG. If a system is represented by the general equation

$$aA + bB \longrightarrow cC + dD$$

then the expression for the change in free energy is

$$\Delta G = \Delta G^\circ + RT \ln \frac{[C]^c[D]^d}{[A]^a[B]^b} \qquad (1)$$

where ΔG° represents a standard free energy change, T is the absolute temperature, R is the gas constant, ln is the natural log, and the bracketed quantities are the activities** of the species.

Another expression for G is:

$$\Delta G = \Delta H - T\Delta S \qquad (2)$$

ΔH is the enthalpy change of the reaction and ΔS is the entropy change.

At constant pressure:

$$\Delta H = \Delta E + P\Delta V \qquad (3)$$

Substituting (3) into (2)

$$\Delta G = \Delta E + P\Delta V - T\Delta S \qquad (4)$$

By the First Law of Thermodynamics:

$$\Delta E = q - w \qquad (5)$$

q = heat **added** to the system

w = work done **by** the system

* Named for J.W. Gibbs. See introduction to Experiment 23 on thermodynamics.

** The activity is the effective concentration. The equilibrium constant expression holds only for ideal solutions. When the concentrations are low, the activity is nearly the same as the concentration.

The Second Law of Thermodynamics says

$$\Delta S = \frac{q_{rev}}{T} \qquad (6)$$

q_{rev} = heat added in a process carried out reversibly

Substitution of (5) and (6) into (4) yields

$$\Delta G = q - w + P\Delta V - q_{rev} \qquad (7)$$

Rearranging

$$\Delta G = (q - q_{rev}) - w + P\Delta V \qquad (8)$$

In an electrochemical cell there are two kinds of work. One is the work of expansion ($P\Delta V$). The second is the electrical work done by the cell (w_{elect}). Substituting $P\Delta V$ and w_{elect} for w in equation (8) we obtain

$$\Delta G = (q - q_{rev}) - (P\Delta V + w_{elect}) + P\Delta V \qquad (9)$$

Simplifying yields

$$\Delta G = (q - q_{rev}) - w_{elect} \qquad (10)$$

If the electrical process is carried out reversibly, then equation (10) reduces to

$$\Delta G = -w_{elect} \qquad (11)$$

The free energy change depends on the electrical work done by the system.

The symbol $\mathcal{E}$ is used to express the potential difference between electrodes. When charges move from one electrode to another, they can do work if they move spontaneously (galvanic cell). If they do not move spontaneously, work must be done to move them (the electrolytic cell). The potential difference $\mathcal{E}$ (which can be expressed in volts) is work (joule) per unit charge (coulomb).

$$\mathcal{E} = \frac{w}{Q} \quad \text{and} \quad w = \mathcal{E}Q$$

where Q is expressed in coulombs, $\mathcal{E}$ is the potential difference in volts, and w is work in joules.

If nF is the total charge (Q), and $\mathcal{E}$ is measured in volts, then

$$w = nF\mathcal{E} \qquad (12)$$

where n is moles of electrons and F is the faraday. This is an equation for electrical work.

When electrons move from anode to cathode in the external circuit, the amount of work which the system can do depends upon the number of electrons passed, the number of faradays, and the potential drop (volts).

If $\Delta G = -w = -nF\mathcal{E}_{cell}$ then for ΔG to be negative and indicate a spontaneous process, the $\mathcal{E}_{cell}$ must be positive.

$$\Delta G = -nF\mathcal{E}_{cell} \qquad (13)$$

$$\Delta G° = -nF\mathcal{E}°_{cell} \qquad (14)$$

$\mathcal{E}°_{cell}$ is the standard potential of a cell. This is the algebraic sum of the standard potentials of the half-cells compared to a standard hydrogen electrode (arbitrarily at zero potential) in which the activities of the components are 1. Substituting Eqs. (13) and (14) in Eq. (1)

$$-nF\mathcal{E}_{cell} = -nF\mathcal{E}^{o}_{cell} + RT \ln \frac{[C]^{c}[D]^{d}}{[A]^{a}[B]^{b}}$$

Dividing by nF and multiplying by -1

$$+\mathcal{E}_{cell} = +\mathcal{E}^{o}_{cell} - \frac{RT}{nF} \ln \frac{[C]^{c}[D]^{d}}{[A]^{a}[B]^{b}}$$

(Nernst equation)

If the temperature is 298°K, R is 8.314 joule*/°K mole, F is 96,493 coulombs, and ln = 2.303 log, then

$$\mathcal{E}_{cell} = \mathcal{E}^{o}_{cell} - \frac{0.059}{n} \log_{10} \frac{[C]^{c}[D]^{d}}{[A]^{a}[B]^{b}}$$

For most calculations in dilute solutions, the values for the activities are approximately equal to the values for the concentration, and pure liquids and pure solids are arbitrarily assigned the value of 1.

pH Titrations

For a pH titration the electrode in common use is the glass electrode, which is a tube fitted with a special conducting glass membrane. If the pH inside the electrode is different from the pH outside of the electrode, a potential difference inside and outside of the electrode is established. The changes in potential as the pH changes can be measured.

The glass electrode is built with an internal reference electrode which is immersed in an acid solution of fixed hydrogen ion concentration. By immersing this electrode into a solution of different hydrogen ion concentration, a potential difference occurs at the thin glass membrane which separates the two solutions. Then by using a reference electrode, such as a calomel electrode, it is possible to measure the potential difference between the two electrodes. The glass of the membrane is able to exchange protons in the solution for sodium ions in the glass. The amount of exchange decreases as the membrane is penetrated more deeply. The penetration of protons occurs on both sides of the membrane and a "dry" glass layer remains between both sides. Current is carried through the membrane by the transfer of both protons and sodium ions where the glass is hydrated and through the dry glass region by sodium ions. Use of the glass electrode in conjunction with a pH meter requires that the electrode be immersed in a buffer of known hydrogen ion concentration for a period of time until the drift in observed pH values stops and the meter can be adjusted for that particular pH. The buffer solution selected for this meter calibration should have approximately the pH expected at the endpoint of the analysis.

The relationship between potentiometric (voltage) readings and pH is directly proportional for the following reason.

The Nernst equation would give

$$\mathcal{E}_{cell} = \mathcal{E}_{ref} + 0.059 \log_{10} \frac{1}{[H^{+}]}$$

* A joule is also equivalent to a volt-coulomb.

if all other species are at unit activity. Therefore, the potential of the cell depends upon the concentration of the hydrogen ion. Since

$$pH = \log \frac{1}{[H^+]} = -\log [H^+]$$

$$pH = \frac{\mathcal{E}_{cell} - \mathcal{E}_{ref}}{0.059}$$

To calculate pK_a consider the following:
The ionization constant expression may be developed from the equation for the dissociation of a weak acid.

$$HA \rightleftarrows H^+ + A^-$$

where HA is any weak acid and A^- is the anion of the acid.

$$\frac{[H^+][A^-]}{[HA]} = K_a$$

The square brackets denote that the concentration of the enclosed species is expressed in moles/liter.

Solving this equation for $[H^+]$ gives

$$[H^+] = K_a \frac{[HA]}{[A^-]}$$

When the concentrations of $[A^-]$ and $[HA]$ are equal, then $[H^+]$ equals K_a of the acid.

$$[H^+] = K_a$$

pH is $-\log [H^+]$ and pK is defined as $-\log K_a$. Therefore

$$pH = pK_a$$

when $[HA] = [A^-]$.

Volumetric analyses make use of the fact that irrespective of the type of reaction taking place,* the greatest change in concentration occurs in the vicinity of the equivalence point. When pH is changing, as in acid-base titrations, the greatest change in pH occurs at the equivalence point.**

The changes in pH in this experiment will be monitored electrically. It is also possible to detect the endpoint using visual indicators. These visual indicators are complex organic weak acids or bases whose molecular and ionic forms have colors radically different from each other. The particular color predominating in a solution containing a very small amount of indicator depends on the pH of the solution. If the visual indicator is carefully chosen, there is a distinct change in the color of the solution at or very near the equivalence point of the titration because the pH change is so marked. It can be shown that an effective indicator is one whose pK is within plus or minus one unit of the pH at the equivalence point of the titration.

* These reactions may be acid-base, oxidation-reduction, precipitation, complex formation.

** The equivalence point is achieved when stoichiometric quantities of base exactly neutralize the acid. When this point is determined experimentally by a color change, electrically, or by some other change in the property of the solution, it is called the endpoint. Ideally the endpoint and the equivalence point coincide.

INVESTIGATION

Purpose: To determine the gram equivalent weight and pKa of a weak acid by titration with a pH meter; to observe how well the indicator endpoint correlates with the potentiometric endpoint.

Equipment:

- 1 50-ml buret
- 2 250-ml beakers
- 1 25-ml transfer pipet
- 2 250-ml volumetric flask
- 1 wash bottle
- 1 500-ml Florence flask
- 1 pH meter
- 1 electrode support
- 1 glass electrode (Ag/AgCl triple)
- 1 calomel reference electrode
- 1 automatic stirrer
- 1 ring stand
- 1 buret clamp
- 3 150-ml beakers

Chemicals:

oxalic acid, $H_2C_2O_4 \cdot 2H_2O$, solid
sodium hydroxide, NaOH, solid
phenolphthalein, 0.1% solution
potassium acid phthalate, $KHC_8H_4O_4$, solid

Procedure:

I. Preparation of the Sodium Hydroxide Solution

Calculate the amount of sodium hydroxide you will need to prepare 500 ml of a 0.5 M solution. Weigh about this amount into a weighed **150-ml** beaker. Transfer the sodium hydroxide to a **500-ml** Florence flask and fill to the neck with deionized water. Invert the flask several times to dissolve the base. **What is the approximate normality of this solution?**

II. Preparation of the Oxalic Acid Solution

Calculate the amount of oxalic acid required to prepare 250 ml of a 0.25 M solution. Weigh about this amount precisely into a **150-ml** weighed beaker. Transfer the oxalic acid quantitatively to a **250-ml** volumetric flask. It is important not to lose any of this acid because this is your standard, and the accuracy of your results will depend upon the care taken in the preparation of this solution.

Remove any crystals which adhere to the beaker with a gentle stream of deionized water from your wash bottle. When all the oxalic acid has been transferred and the flask is about one-half full, close the flask tightly and invert several times to put the acid into solution. When the acid is completely dissolved, add deionized water to the mark and close the flask.

What is the normality of this oxalic acid solution?

Transfer this solution to a *clean dry* flask, properly labeled with solution concentration.

III. The Calibration of the PH Meter

The pH meter is calibrated with a buffered solution having a pH in the range of 7 to 9 because this is approximately the region where the endpoint will occur. Using the precise value of the pH of the buffered solution for calibration of the meter, consult your instructor on how to connect the electrodes to the pH meter and set the temperature control.

Pour about 120 ml of the solution into a 150-ml beaker and carefully immerse the glass and calomel electrodes. Turn the pH meter on. Allow the electrodes to remain in the solution while you are preparing the other solutions.

When the pH meter reads a constant value, set it for the pH of the buffer solution. Put the meter on standby. Remove the buffer solution, and prepare to start the titration.

IV. Standardization of Base

Into a 250-ml beaker pipet exactly 25 ml of standard oxalic acid. Then ad enough water to fill half the beaker, 3 drops of phenolphthalein indicator, and the magnetic stirrer. Immerse the electrodes in the solution. *Be sure there i clearance between the bottom of the electrodes and the magnetic stirrer. Have this checked by your instructor* **(Fig. 14)**.

Scrub your 50-ml buret thoroughly with detergent and water. Rinse with ta water, and then with 5-ml increments of **deionized water.** Follow this with thre rinsings of 5-ml portions of 0.5 M sodium hydroxide solution. Discard all rins ings. Fill the buret with the base. *Avoid splashing the solution as it is poured into the buret.* Be sure to eliminate air bubbles from the stopcock area Drain the sodium hydroxide solution to just below the 0.00 mark. Support the buret with a clamp so that the tip is about 3 cm above the solution in the 250-ml beaker. Turn on the magnetic stirrer and the pH meter. Record the pH of th acid solution to the nearest 0.1 pH unit and the buret reading to the nearest 0.03 ml in the table on the data page. **(IVa.)**

a. Titration of sample 1

Add about 1 ml of 0.5 M sodium hydroxide. Record the buret reading; then the pH reading. Continue the titration in 1 ml increments, recording buret rea ings and pH (columns 1 and 2 of Table **IVa**) until you have passed the endpoint. This is very obvious from both the color change of the indicator and the large change in pH.

b. Titration of sample 2

From the first titration the endpoint is located to within 1 ml of your base. For this second titration, refill your buret with the sodium hydroxide, again pipet exactly 25 ml of oxalic acid into a clean 250-ml beaker, add **deionized** water to fill half the beaker, 3 drops of phenolphthalein, and the magnetic stirrer. Insert the electrodes and recheck their position above the stirrer.

Perform the second titration by first recording the volume of base in the buret and the pH of the acid solution. Add the sodium hydroxide solution in approximately one ml increments until you are one ml short of the endpoint. Then proceed dropwise. Record both the buret reading and the pH until it appears that you are one ml past the endpoint. You will observe the endpoint not only by the large change in pH but also by the change in the indicator color. Observe whether both endpoints occur simultaneously.

There are several graphical methods for determining the endpoint from the data you have recorded. For the determination of the normality of the sodium hydroxide prepare a differential plot. You will graph the change in pH per unit volume of titrant added against the volume of titrant added. The volume of titrant at which the value of $\Delta pH/\Delta v$ is the greatest represents the endpoint.

Explicit instructions are given in the data pages.

V. Determination of the Gram Equivalent Weight of an Unknown Acid*

Obtain an unknown acid and record its number on the data page.

a. Weigh precisely a sample of unknown acid** and dissolve it in deionized water and quantitatively transfer it to a 250-ml beaker. Add 3 drops of phenolphthalein. Insert the electrodes and the magnetic stirrer. (Check to be sure electrodes are positioned above the stirrer.) Add enough deionized water to immerse the electrodes. Clamp the buret filled with standardized NaOH above the beaker. Be sure there are no trapped air bubbles in the stopcock area. Record buret and pH readings.

b. Titrate the first sample in 1 ml increments. Record both pH and buret readings. Continue the titration 2 ml past the endpoint. The endpoint is known now within 1 ml for this sample.

Calculate the ratio for the number of milliliters of base per gram of acid required for the approximate endpoint. The volume of base used in the calculation is the total number of milliliters required to be *just short* of the endpoint.

$$\frac{\text{ml of base for titration 1}}{\text{g of sample 1}}$$

Using this approximation, calculate the volume of base required for the second sample. Your calculations should look like this:

$$\frac{(\text{ml of base for titration 1})}{(\text{g of sample 1})}\ (\text{g of sample 2}) = \text{ml for sample 2}$$

c. Weigh a second sample of the unknown acid and dissolve as described in (a).
Perform the second titration by first recording the volume of base in the buret and the pH of the acid solution. Add the sodium hydroxide in approximate-

---

* A discussion of equivalent weight can be found on page 60.

** The weight of unknown required for a titration of 20-30 ml base can be determined by trying one gram of acid and titrating with the base for quick approximation. If time is short, this information may be given by the instructor.

Figure 14. An arrangement of electrodes, buret, and stirrer in operation.

ly one ml increments until you are one ml short of the endpoint. Then proceed dropwise. Record both the buret reading and the pH until the endpoint is passe

Determine the endpoint for the titration by preparing a differential plot similar to the one graphed for the sodium hydroxide standardization.

Calculate the gram equivalent weight of the 'unknown' acid.

i. Did the indicator change color at the endpoint?

VI. Determination of pK_a for the 'Unknown' Acid

For this determination, another type of graphical analysis will be used.

Make a graph of the pH on the y-axis vs. ml of sodium hydroxide on the x-axis using the data from the second titration.

Draw a line from the point of steepest slope of the curve to the x-axis. This gives the exact amount of sodium hydroxide required to neutralize the acid.

Divide this volume in half. You now know how much base is required to neu tralize one-half the acid present. Locate this half-neutralization point on th titration curve. A line drawn from this point parallel to the x-axis and inter secting the y-axis gives the pH when the concentration of the undissociated aci equals the concentration of the anion. This pH is equal to the numerical value of the pK_a of the weak acid.

If there is more than one replaceable hydrogen in the acid, and if the pK' for the separate ionizations are close in value, then the experimentally determined pK is the average of the sum of the individual pKs.

NAME____SECTION_____________DATE__________

ACID-BASE TITRATION USING THE pH METER
DETERMINATION OF THE pKa OF A WEAK ACID

DATA

I. Preparation of the Sodium Hydroxide Solution

a. amount of NaOH needed to prepare 500 ml of a 0.5 M solution ____________g

b. weight of 150 ml beaker and NaOH ____________g

weight of 150 ml beaker ____________g

weight of NaOH ____________g

normality of NaOH (approximate) ____________

II. Preparation of the Standard Oxalic Acid Solution

a. calculated amount of $H_2C_2O_4 \cdot 2H_2O$ required to prepare 250 ml of a 0.25 M solution ____________g

b. weight of 150 ml beaker and $H_2C_2O_4 \cdot 2H_2O$ ____________g

weight of 150 ml beaker ____________g

weight of $H_2C_2O_4 \cdot 2H_2O$ ____________g

molarity of oxalic acid ____________g

normality of oxalic acid ____________g

III. Standardization of the Sodium Hydroxide

a. Titration of sample 1

Buret reading	pH	Total NaOH (ml)
		0

Bracket the reading of greatest pH change.

NAME ______________________________ SECTION ____________ DATE ____________

III. b. Titration of sample 2

1	2	3	4	5	6
Buret reading	pH	Total NaOH (ml)	ΔpH	Δv	$\frac{\Delta pH}{\Delta v}$
		0.00	0	0	0

Indicator endpoint ____________

Graphically determined endpoint ____________
(see the following instructions for this calculation below)

III. C. Calculations for the Preparation of the Differential Plot

A. Complete the preceding table, filling in columns 3-6. Note that:

1. The milliliters of NaOH, column 3 in the table, is the buret reading minus the initial buret reading

2. ΔpH, column 4, is found by calculating the change in pH for each successive increment of NaOH.

3. Δv, column 5, is the value for the exact volume of each increment of base. In titration 2 there should be about 45 entries.

4. ΔpH/Δv, column 6, is the value of ΔpH entered in column 4 divided by the value of Δv entered in column 5.

B. Make a graph of the results of titration 2 by plotting ΔpH/Δv on the y-axis and volume of NaOH added on the x-axis.

C. Using the results of this plot, calculate the exact normality of the NaOH solution. Show calculations.

Normality of NaOH ____________________

IV. Determination of the Equivalent Weight of an 'Unknown' Acid

a. weight of sample	1	2
weight of beaker + unknown	_____g	_____g
weight of beaker	_____g	_____g
weight of unknown	_____g	_____g

Unknown Acid Number ______________

NAME____SECTION______________DATE____________

IV. b. Titration of sample 1 of unknown

Buret reading	pH	Total NaOH (ml)
		0.00

Calculation of ml base/g sample

IV. c. Titration of sample 2 of unknown

Buret reading	pH	Total NaOH (ml)	ΔpH	Δv	ΔpH/Δv

Volume of NaOH required to neutralize unknown acid sample (determined graphically) ________ ml

d. Calculation of the equivalent weight of the unknown acid

$$\text{equivalent weight} = \frac{(\text{g of sample})}{(\text{ml of base})}\,\frac{(1)}{(N_{base})}\,\frac{(10)^3\ \text{ml}}{(\text{liter})} =$$

$$= \frac{\text{g}}{\text{ml}}\,\frac{(1)}{(\text{eq/l})}\,\frac{(10^3\text{ml})}{(\text{liter})} = \text{g/eq}$$

Calculated gram equivalent weight = ________________

Show calculations.

i. Did the indicator change at the endpoint?

NAME____________________ SECTION__________ DATE__________

V. Determination of the pK_a of the Unknown Acid

a. 1. Make a graph of the pH vs. ml of NaOH

2. Find the half-neutralization point. __________ml

3. From the half-neutralization point, draw a line parallel to the x-axis intersecting the y-axis. What is the pH value at this point? __________

b. What is the pK_a of the acid? __________

What is K_a of the acid? (Show method of calculations)__________

THOUGHT

1. Why is it necessary to calibrate the pH meter before titrating?

2. Methyl yellow changes from red to yellow when the pH increases from 2.9 to 4.0. Could this indicator have been used to signal the endpoint in these titrations? Explain briefly.

3. Compare the volume of base required for neutralization of the acid determined from the plot in section V c with that determined from the graph in section VI.

NAME____SECTION__________DATE__________

Experiment 14

TITRATION: OXIDATION-REDUCTION

PRELIMINARY QUESTIONS

1. a) What is an oxidizing agent?

   b) What is a reducing agent?

   c) What is meant by oxidation number?

2. a) In determining the normality of a reagent why is it necessary to specify the reaction?

   b) In determining the molarity of the reagent why is it *unnecessary* to specify the reaction?

3. What is: a) the molecular weight of KI?

   b) the gram equivalent weight of KI in a reaction in which the KI is converted to I_2?

c) If 10 g of KI are dissolved in 100 ml of solution, what is the molarity of the KI?

d) What is the normality of the KI in (c) if it is to be converted to I_2 in a redox reaction?

4. a) What is the gram equivalent weight of $H_2C_2O_4 \cdot 2H_2O$ in a redox reaction in which CO_2 is formed?

b) What weight of $H_2C_2O_4 \cdot 2H_2O$ is necessary to prepare 250 ml of a 0.1 N solution for the reaction in (a)?

c) What is the molarity of the oxalic acid solution?

5. a) What is the gram equivalent weight of $KMnO_4$ in a reaction in acid solution in which Mn^{2+} is formed?

b) What weight of $KMnO_4$ is required to prepare 500 ml of 0.1 N solution for the reaction in (a) above?

c) What is the molarity of this solution?

NAME________________________ SECTION____________ DATE______________

6. Using half reactions balance the net ionic equations for the reaction of $KMnO_4$ with $H_2C_2O_4$ in acid solution:

$$MnO_4^- + H_2C_2O_4 + H^+ \longrightarrow H_2O + CO_2 + Mn^{2+}$$

7. a) Calculate the number of equivalents of $KMnO_4$ contained in 25 ml of a 0.14 N solution.

   b) How many equivalents of $H_2C_2O_4$ will be necessary to react completely with this amount of $KMnO_4$?

   c) What volume of 0.10 N $H_2C_2O_4$ will be required to react completely with this amount of $KMnO_4$?

8. a) One-half gram of a reducing agent is dissolved in 50 ml of water. Titration of this solution requires 26.52 ml of 0.10 N $KMnO_4$ solution. How many equivalents of $KMnO_4$ are used?

   b) What is the gram equivalent weight of the unknown reducing agent?

Experiment 14

TITRATION: OXIDATION-REDUCTION

IDEAS

The words oxidation and reduction were not always paired. Reduction appears to have been used in the early period of chemistry to indicate the preparation of a metal from its ore or its "calx" (the substance obtained from heating a metal in air), because the resulting metal weighed less than the substance from which it was made. The preparation of calxes (calcination) and metals (reduction) and the phenomena of burning and respiration were all explained by the phlogiston theory. Proposed in the seventeenth century, the phlogiston theory was one of the early great generalizations in chemistry, relating the phenomena of calcination, reduction, burning, and respiration in terms of fluid "phlogiston." In summary, the theory assumed the liberation of phlogiston during combustion, calcination, and respiration, the presence of air being necessary to absorb phlogiston. The phlogiston theory may be illustrated as follows:

lead calx + phlogiston = lead

lead (heated in air) $\longrightarrow$ lead calx + phlogiston

charcoal is almost pure phlogiston

lead calx + charcoal $\longrightarrow$ lead

The reduction of lead calx to lead was therefore accomplished by the addition of phlogiston. Why the lead weighed less than the calx (despite the addition of phlogiston to the calx) posed a problem which was ignored by some. Others suggested the additional assumption that phlogiston could have negative weight. This anomaly did not cause the overthrow of the phlogiston theory; as is usual in the history of science, an entrenched theory is overthrown only after a new and better theory has been formulated.

A new theory of burning was suggested by A. Lavoisier (1743-1749), and it eventually replaced the phlogiston theory. Lavoisier proposed that during combustion, calcination, and respiration, something was taken from the air (rather than given to the air). This he called oxygen, from the Greek "oxygine principle," meaning "acidifying principle," because he mistakenly thought that it is contained in all acids. Oxygen could be released from its union with metals in compounds during reduction. The word "oxidation" has its roots in the discovery by Lavoisier of the role of oxygen in various chemical processes, for oxidation at first was used to describe chemical union with oxygen. During Lavoisier's early attempts to publicize his theory of combustion, he wrote, "I do not expect that my ideas will be adopted all at once; human nature bends towards one viewpoint, and those who have envisaged nature from a certain point of view during a part of their career, change only with difficulty to new ideas; it is time, then, to confirm or destroy the opinions which I have presented."*

The modern usage of oxidation and reduction followed late-nineteenth century investigations that led to the idea that atoms are not solid, indivisible spheres, as Dalton had hypothesized, but have internal structure. The discovery of the electron, and the hypothesis by Rutherford of the nuclear atom with revolving electrons, followed.

In an oxidation-reduction titration, the essential reaction is a transfer of electrons from one species to another. This may be compared to an acid-base reaction in which there is a transfer of protons from one reagent to another. The chemical which gives up the electrons is said to be oxidized and the other species, which gains the electrons is said to be reduced. The use of oxidation numbers is an arbitrary but universally accepted method for determining which reagents are oxidized and which are reduced. It also simplifies the balancing of redox** reactions, some of which can be quite complicated. For a complete discussion of oxidation numbers and the balancing of redox equations, refer to a textbook.

In redox reactions there must be an oxidizing agent and a reducing agent.

A. Reducing Agents

The reagent which gives up electrons (i.e., is oxidized) is called the reducing agent. Some common reducing agents are zinc metal, oxalic acid ($H_2C_2O_4$) and iodide ion (I^-).

a) $Zn \rightarrow Zn^{2+} + 2e$

b) $I^- \rightarrow 1/2\, I_2 + e^-$

c) $H_2C_2O_4 \rightarrow 2CO_2 + 2H^+ + 2e^-$

The above equations are called half-reactions. They show the changes that only one of the reagents in a redox reaction undergoes.

One equivalent of a reducing agent is that quantity of reagent which gives up one mole of electrons.

The gram equivalent weight of a reducing agent is that weight which furnishes one mole of electrons.

Referring to the half-reactions above, you see that one mole of zinc furnishes 2 moles of electrons. Thus the equivalent weight of zinc is

$$\frac{(65.38\text{ g})}{(\text{mole Zn})} \quad \frac{(1\text{ mole Zn})}{(2\text{ moles electrons})} \quad \frac{(1\text{ mole electrons})}{(\text{eq})} = \frac{32.69\text{ g}}{(\text{eq})}$$

One mole of I^- furnishes one mole of electrons, so that if the iodide were present in solution as KI, the equivalent weight of the KI would be the same as the molecular weight.

To summarize these relationships:

$$\text{gram equivalent weight of a reducing agent} = \frac{\text{gram molecular wt}}{\dfrac{\text{moles of electrons lost}}{\text{mole of reagent}}} = \frac{\dfrac{\text{g}}{\text{mole}}}{\dfrac{\text{eq}}{\text{mole}}} = \frac{\text{g}}{\text{eq}}$$

* F. Bonner and M. Phillips, *Principles of Physical Science*, Addison-Wesley Publishing Company, Reading, Mass., 1957, p. 127.

** Redox is a useful abbreviation for oxidation-reduction.

$$\frac{\text{number of equivalents}}{\text{of reducing agent}} = \frac{\text{wt of reducing agent}}{\text{gram equivalent wt}} = \frac{g}{g/eq} = eq$$

B. The Oxidizing Agent

The reagent which takes on electrons is called the oxidizing agent. It is reduced (i.e., there is a reduction in the value of the oxidation number). Typical oxidizing reagents are MnO_4^- (permanganate ion), $Cr_2O_7^{2-}$ (dichromate ion) and NO_3^- (nitrate ion).

Reactions for these chemicals in acid solution may be represented as:

a) $MnO_4^- + 8H^+ + 5e^- \rightarrow Mn^{2+} + 4H_2O$

b) $Cr_2O_7^{2-} + 14H^+ + 6e^- \rightarrow 2Cr^{3+} + 7H_2O$

c) $NO_3^- + 4H^+ + 3e^- \rightarrow NO + 2H_2O$

One equivalent of an oxidizing agent is that weight which gains one mole of electrons. For an oxidizing agent:

$$\text{gram equivalent weight} = \frac{\text{gram molecular weight}}{\dfrac{\text{number of moles electrons gained}}{\text{mole of reagent}}} = \frac{\dfrac{g}{\text{mole}}}{\dfrac{eq}{\text{mole}}} = \frac{g}{eq}$$

$$\frac{\text{number of equivalents}}{\text{of oxidizing agent}} = \frac{\text{wt of oxidizing agent}}{\text{gram equivalent wt}} = \frac{g}{g/eq} = eq$$

If the solution of $Cr_2O_7^{2-}$ is prepared from $K_2Cr_2O_7$ then the equivalent weight is one-sixth the molecular weight of potassium dichromate, since only one-sixth of a mole is needed to provide one mole of electrons.

Normality is defined as the number of equivalents of solute per liter of solution. In these reactions it is the number of moles of electrons gained or lost per mole of reactant that determines the number of equivalents of each reagent.

$$N = \text{normality} = \frac{\text{number of equivalents of redox agent}}{\text{volume in liters}}$$

C. The Redox System

Each of the half-reactions for oxidation must be coupled with a half-reaction for reduction. The equations can be balanced by the ion-electron method which has the advantage of indicating the species as they exist in solution.

In a redox titration, the principle involved is that:

$$\text{number of equivalents of reducing agent} = \text{number of equivalents of oxidizing agent}$$

$$\text{number of electrons lost} = \text{number of electrons gained}$$

and

$$N_R V_R = N_o V_o$$

Normality and molarity are related as follows:

$$\frac{\text{number of eq}}{\text{liter}} = N = (M)\,\frac{(\text{number of eq})}{(\text{mole})} = \frac{(\text{number of moles})}{(\text{liter})}\,\frac{(\text{number of eq})}{(\text{mole})}$$

The value for the molarity of a solution is fixed, and depends only on the weight of the solute in the particular volume and on the molecular weight of the reagent which is a constant value. However, it should be apparent that this is not true for normality. Nitrate ion can be reduced to several different products depending on the reducing reagents used and on the conditions of the experiment.

$$NO_3^- + 4H^+ + 3e \rightarrow NO + H_2O$$

$$NO_3^- + 2H^+ + e^- \rightarrow NO_2 + H_2O$$

In the first reaction there are three equivalents per mole of nitrate ion and in the second reaction there is only one equivalent per mole of nitrate ion. A 0.1 M nitrate solution is 0.3 N if the first reaction occurs and 0.1 N in the second reaction. It is therefore very important when making up known solutions for redox titrations to know what the products of the reaction are so that the proper weights of reagents are used.

This experiment presents a method for titrating a solution of a reducing agent, oxalic acid, with potassium permanganate solution, a strong oxidizing agent. Potassium permanganate forms a deep purple solution which becomes colorless when complete reduction to manganous ion (Mn^{2+}) has occurred. This color change is very sensitive, so that in permanganate titrations the endpoint can be detected without the use of an indicator.

Once the potassium permanganate has been standardized, the equivalent weight of an unknown reducing agent will be determined. The relationships which permit calculation of this equivalent weight are similar to those used in determining the equivalent weight of an unknown acid. (See Experiment 17.)

$$\text{number of eq reducing agent} = \text{number of eq oxidizing agent}$$

$$\frac{\text{wt reducing agent}}{\text{g eq wt reducing agent}} = \frac{\text{g}}{\frac{\text{g}}{\text{eq}}} = N_o V_o = \frac{\text{eq}}{\text{liter}}\,(\text{liters})$$

INVESTIGATION

Purpose: To determine the gram equivalent weight of an unknown reducing agent.

Equipment:

- 1 50-ml buret
- 1 25-ml transfer pipet
- 1 50-ml transfer pipet
- buret clamps
- 1 250-ml beaker
- 1 500-ml Florence flask
- 1 250-ml volumetric flask

Chemicals: potassium permanganate, $KMnO_4$, solid
oxalic acid, $H_2C_2O_4 \cdot 2H_2O$
6 M sulfuric acid, H_2SO_4

Procedure:

I. Preparation of Potassium Permanganate Solution

The equivalent weight of an unknown reducing agent will be found by titrating it against a standard potassium permanganate solution. Because of impurities present in solid potassium permanganate, a standard solution of potassium permanganate cannot be prepared directly. To prepare a standard permanganate solution, make up an approximately 0.1 N solution, and then determine its correct normality by titrating it against a standard oxalic acid solution.

Calculate the amount of potassium permanganate needed to make up 500 ml of 0.1 N solution. Weigh an amount close to this calculated quantity into a 250-ml beaker. Add to this 150 ml of the total 500 ml of deionized water. Boil this solution 10 to 15 minutes. Cool the solution and let it stand for about 30 minutes. Then carefully decant the supernatant liquid into a Florence flask and add the remaining 350 ml of water. Close the flask with a rubber stopper and label it with the name of the contents, concentration, the date, and your initials.

II. Preparation of Standard Oxalic Acid Solution

Calculate how much oxalic acid is required to prepare 250 ml of a 0.1 N solution when the oxalate is oxidized to CO_2. Include the water of hydration in determining the molecular weight. The formula is $H_2C_2O_4 \cdot 2H_2O$.

Weigh precisely an amount which is close to the calculated value and record the weight. *It is not necessary to spend time trying to weigh out your exact calculated value.*

Record all data on data pages.

Add a small amount of deionized water to the oxalic acid crystals you have weighed and transfer them with care into the volumetric flask. Use a stream of deionized water from your wash bottle to remove any trace crystals that may have adhered to the glass container, adding these washings to your volumetric flask. Add enough water so that your flask is about one-half full. Stopper the flask and agitate the contents until complete solution has been achieved. Oxalic acid is not very soluble, and it may take some time to dissolve completely.

Fill the flask to just *below* the mark. Bring the solution up to the mark by adding the last few drops with a medicine dropper or a wash bottle. Stopper your volumetric flask tightly so that no solution will be lost before it is completely mixed. Mix the contents of the flask by inverting it about 25 times, until no streaming lines can be observed. Your solution is now ready to be used as a standard. Since its concentration is known, you must be careful not to do anything to change it.

The normality of your solution can be calculated from the following:

$$\text{number of equivalents of oxalic acid} = \frac{\text{grams of oxalic acid}}{\text{g eq wt}}$$

$$N = \frac{\text{number of equivalents of oxalic acid}}{\text{liters of solution}}$$

III. Standardization of the Potassium Permanganate Solution

Using the 25-ml transfer pipet, add 25 ml of oxalic acid to a clean 250-ml Erlenmeyer flask. Add 50 ml of deionized water and 15 ml of 6 M sulfuric acid. Heat this solution to 80°C and titrate with your potassium permanganate solutio to a pink endpoint which lasts 15 seconds. The reduction of potassium permanganate is initially quite slow. The reaction is autocatalyzed by manganese (II) ion, and once some of it has formed, the reaction rate will accelerate. The endpoint is sharp only at elevated temperatures. Hence the need for titrating a hot solution.

i. What purpose does the sulfuric acid serve?

ii. Is it necessary to measure the sulfuric acid quantitatively?

iii. Using the balanced equation and the quantities of oxidizing and reducing agent used in your titration, calculate the number of equivalents of sulfuric acid required for the reaction.

iv. How many equivalents did you add?

v. How many equivalents of acid are in excess?

vi. Can you explain what effect this excess acid will have on the reaction in terms of Le Chatelier's principle? (See your textbook)

Titrate two more 25-ml samples of 0.1 N oxalic acid. Record all your data

Calculate the exact normality of the potassium permanganate solution, keep ing in mind that at the endpoint the number of equivalents of oxalic acid will be equal to the number of equivalents of potassium permanganate.

When you have obtained reproducible results for the normality of your potassium permanganate solution, obtain an unknown from your instructor for the determination of its equivalent weight.

IV. Determination of the Gram Equivalent Weight of an Unknown Reducing Agent

In an Erlenmeyer flask, weigh precisely an amount of an unknown suggeste by your instructor. Dissolve the unknown in 50 ml of deionized water, add 15 ml of 6 M sulfuric acid, and heat. Titrate the sample at 80°C with your standard potassium permanganate solution. Perform the analysis three times. Record all data on the data pages.

NAME______________________________ SECTION__________ DATE__________

Unknown number __________

TITRATION: OXIDATION-REDUCTION

DATA

I. Weight of beaker and potassium permanganate __________

Weight of beaker __________

Weight of potassium permanganate __________

II. Calculation for the preparation of 250 ml of 0.1 N oxalic acid solution using $H_2C_2O_4 \cdot 2H_2O$

Weight of beaker and oxalic acid __________

Weight of beaker __________

Weight of oxalic acid __________

Normality of oxalic acid: (Show method of calculation)

III. i. What purpose does the sulfuric acid serve?

ii. Is it necessary to measure the sulfuric acid quantitatively? ______

iii. How many equivalents of sulfuric acid are required for the reaction? (Show method of calculation)

____

iv. How many equivalents did you add? ___________
(Show method of calculation)

v. How many equivalents of acid are in excess? ___________
(Show method of calculation)

vi. Can you explain what effect this will have on the reaction in terms of Le Chatelier's principle?

IV. Standardization of the potassium permanganate

Reactants	Trials I	II	III
ml of $H_2C_2O_4$			
Normality $H_2C_2O_4$			
Number of equivalents $H_2C_2O_4$			
Buret reading, final for $KMnO_4$			
Buret reading, initial for $KMnO_4$			
Volume of $KMnO_4$			
Calculated normality of $KMnO_4$			
Average Normality			

NAME________________ SECTION__________ DATE__________

IV. Determination of the gram equivalent weight of an unknown reducing agent

Reactants	Trials I	II	III
Grams of unknown			
Buret reading, final for $KMnO_4$			
Buret reading, initial for $KMnO_4$			
Volume of $KMnO_4$			
Equivalents of $KMnO_4$ used			
Gram equivalent weight of unknown			

Average gram equivalent weight ____________

Unknown Number ____________

THOUGHT

1. Potassium permanganate solutions are not stable over long periods of time, forming manganese products of lower oxidation state. What effect will the use of a deteriorated solution have on the calculated gram equivalent weight of an unknown reducing agent?

2. Does the amount of water used to dissolve the unknown reducing agent have to be known exactly? Explain.

3. Why is it not necessary to add an indicator in this titration?

4. a. When used in an acid base titration, what is the normality of 0.1 M HNO_3

   b. When used as an oxidizing agent in which NO is formed, what is the normality of 0.1 M HNO_3?

NAME________________________________SECTION____________DATE__________

Experiment 15

ENTHALPY OF REACTION (HESS'S LAW)

PRELIMINARY QUESTIONS

1. Define:

   a) calorie

   b) specific heat

   c) enthalpy change (ΔH)

2. State Hess's law

3. Why is it necessary to indicate the state (s, l, g, aq) of products and reactants when writing a thermochemical equation?

4. Is the heat of neutralization the same for any strong acid reacting with any strong base? Explain and write a net ionic equation.

5. a) If concentrated hydrochloric acid is 12 M, explain how you would prepare 80 ml of a 2 M solution.

   b) If sodium hydroxide pellets are used, explain how you would prepare 80 ml of a 2 M sodium hydroxide solution.

Experiment 15

ENTHALPY OF REACTION (HESS'S LAW)

IDEAS

Both the physical and biological worlds are characterized by a complex interplay of chemical reactions. In all chemical reactions, heat is released or absorbed. Combustion is one of the commonly observed chemical reactions accompanied by the evolution of heat energy, which may be harnessed directly for many purposes, such as the heating of homes, or which may be converted to other forms of energy, such as mechanical energy in the steam engine or the internal combustion engine.

One of the pioneering studies on the heat of chemical reactions was made by Lavoisier and Laplace in 1780. Investigating the heat associated with various reactions, including combustion, they showed that each reaction was associated with a specific amount of heat and that the heat evolved in a chemical reaction is equal to the heat absorbed in the reverse reaction.

Lavoisier had proposed the modern theory of combustion as the combination of a substance with oxygen. He also suggested the hypothesis that respiration is merely slow combustion. With Laplace, Lavoisier attempted to compare the quantity of heat given off by a burning candle with that given off by a live chicken. The quantity of heat was determined by measuring the quantity of ice caused to melt in an insulated container.

Little progress was made in thermochemistry until about fifty years later, when G. Hess (1802-1850) studied heats of reaction in greater detail and stated the law (1840) that bears his name: The heat evolved or absorbed in a chemical reaction is the same whether the process is accomplished in one step or in many. The law suggested, therefore, that both the reaction and the thermodynamic data could be treated algebraically. Although it was an empirical law when Hess proposed it, it was theoretically justified a few years later when the principle of conservation of energy was proposed. Hess's law can be illustrated by the following problem.

Calculate the heat of ionization ($\triangle H_i$) for the following reaction:

1. $HCN_{(aq)} = H^+{}_{(aq)} + CN^-{}_{(aq)} \qquad \triangle H = \triangle H_i$

Use the following information to solve the problem:

2. $HCN_{(aq)} + OH^-{}_{(aq)} = CN^-{}_{(aq)} + H_2O_{(l)} \qquad \triangle H = -2900 \frac{cal}{mole}$

3. $H^+{}_{(aq)} + OH^-{}_{(aq)} = H_2O_{(l)} \qquad \triangle H = -13{,}800 \frac{cal}{mole}$

Arrange the equations (2 and 3) so that their sum will yield equation 1.

$$HCN_{(aq)} + \cancel{OH^-}_{(aq)} = CN^-_{(aq)} + \cancel{H_2O}(l) \qquad \Delta H = -2900 \frac{cal}{mole}$$

$$\cancel{H_2O}(l) = H^+_{(aq)} + \cancel{OH^-}_{(aq)} \qquad \Delta H = +13{,}800 \frac{cal}{mole}$$

$$HCN_{(aq)} = H^+_{(aq)} + CN^-_{(aq)} \qquad \Delta H = 10{,}900 \frac{cal}{mole}$$

The quantity of heat (ΔH) lost or gained by materials depends upon the quantity of substance present (mass), the characteristic capacity of the substance to absorb heat causing a rise in temperature (specific heat), and the change in observed temperature.

$$\Delta H = (m)(s)(\Delta t)$$

m = mass in grams

s = specific heat, cal/g°C

Δt = observed temperature change

When reactions are carried out in a calorimeter to determine heat gained or lost, a portion of that heat is absorbed by the calorimeter and thermometer. Since the reactions in this experiment are carried out in dilute aqueous media, the quantity of heat absorbed by the apparatus is called the water equivalent of the calorimeter: the amount of heat absorbed by the calorimeter divided by the change in temperature of the calorimeter. An example: To 100 g of water at 25°C are added 100 g of water at 45°C. The expected temperature of the mixture would ideally be 35°C. Instead some heat is absorbed by the calorimeter and the temperature of the mixture is only 34°C.

Heat gained by water in calorimeter

$$(100\ g)\left(\frac{1\ cal}{g^oC}\right)(34^oC - 25^oC) = 900\ cal$$

Heat lost by water added to the calorimeter

$$(100\ g)\left(\frac{1\ cal}{g^oC}\right)(45^oC - 34^oC) = 1100\ cal$$

Water equivalent: $\dfrac{200\ cal}{9^oC} = \dfrac{22\ cal}{^oC}$

The water equivalent is then applied to each reaction run in the calorimeter. For example, if the temperature change of a reaction is 6°C, then the amount of heat absorbed by the calorimeter is

$$\frac{(22\ cal)}{(^oC)}(6^oC) = 132\ cal$$

For exothermic reactions, the water equivalent is added to the calculated amount of heat absorbed by the water. For endothermic reactions, heat is withdrawn from the calorimeter and the observed final temperature would be higher than expected. In this case the number of calories of heat extracted from the calorimeter is subtracted from the number of calories of heat lost by the water.

In this experiment you will determine the heats of reaction for two reactions.

1. $H^+ + OH^- \rightarrow H_2O$
2. $Ca + 2H_2O \rightarrow Ca^{2+} + 2OH^- + H_2$

By applying Hess's law you will be able to predict the heat of reaction for a third reaction.

3. $Ca + 2H^+ \rightarrow Ca^{2+} + H_2$

INVESTIGATION

Purpose: To show how Hess's law can be used to determine the molar heat of reaction.

Equipment: 1 styrofoam cup and cover
2 150-ml beakers
2 thermometers (110°C)

Chemicals: newly opened bottle of solid sodium hydroxide, NaOH
concentrated hydrochloric acid, HCl
newly opened bottle of calcium

Procedure:

I. Determination of Water Equivalent of the Calorimeter.*

Place two 100°C thermometers in water which is at room temperature. If they do not read exactly the same, record which thermometer reads higher and by how much (read to the nearest 0.2°C). Subtract this value from subsequent readings with this thermometer.

Measure 80.0 ml water with your graduated cylinder, put it in a styrofoam cup and allow it to come to room temperature. Put a second 80 ml portion in a beaker and heat gently to about 35°C. (Record both temperatures precisely, making a correction if necessary.) Add the warmer water to the calorimeter, cover, and stir gently with your thermometer to mix the solutions. Record the highest temperature value.

i. Would you expect the final temperature of the mixture to be higher or lower than the median value?

ii. Why?

II. Heat of Reaction Between a Strong Acid and a Strong Base.

Prepare 80 ml each of 2 M hydrochloric acid and 2 M sodium hydroxide. Place each solution in a clean, dry 150 ml beaker. Measure the temperatures of both the acid and the base solutions using your calibrated thermometers which must be clean and dry before each measurement. The temperatures of these solutions should be about room temperature and agree with each other to within 0.5°C. When

---

*If a styrofoam cup is used for the calorimeter, then no water equivalent need be calculated because the insulating properties of this material are excellent. Proceed to part II.

equivalent temperatures have been established, pour the base into the calorimete and record its temperature. Then, all at once, pour the acid into the base. Co stir gently with your thermometer and observe the changes in temperature. Recorc the highest temperature.

III. The Heat of Reaction Between Calcium and Water.

Using a newly opened bottle,* weigh accurately 0.5 g of calcium on a watch glass. Measure 100 ml of distilled water into your clean dry calorimeter. Stir the water and record the temperature at 1-minute intervals. When the temperature remains constant, add the calcium all at once to the solution, cover, and stir tc assist reaction. Again observe the changes in temperature, and record the highest temperature attained.

---

*Calcium is oxidized to calcium oxide on exposure to air. Even the sealed bottle contains small amounts of oxidized calcium because of the air present in the bottle.

NAME____SECTION_____________DATE____________

ENTHALPY OF REACTION (HESS'S LAW)

DATA

I. Determination of the water equivalent of the calorimeter

i. Would you expect the final temperature of the mixture to be higher or lower than the median value? __________

ii. Why?

		#1	#2
a.	Thermometer readings at room temperature	____	____
b.	Temperature of water in calorimeter (nearest 0.2°C)	____	
c.	Temperature of water in beaker (nearest 0.2°C)	____	
d.	Temperature of mixture in calorimeter (nearest 0.2°C)	____	
e.	Water equivalent:		

II. The reaction between hydrochloric acid and sodium hydroxide

Assume in the following that the density and specific heat of the solution are the same as that for pure water.

a. weight of 2 M sodium hydroxide solution based on the volume measured and the accepted density _______

b. weight of 2 M hydrochloric acid solution based on the volume measured and the accepted density _______

c. weight of total solution _______

d. initial temperature of reactants 2 M NaOH _____ 2 M HCl _______ to the nearest 0.2° C

e. final temperature of mixture to the nearest 0.2° C _______

f. difference in temperature between reactants and products (Δt) to the nearest 0.2°C (e-d) ______

g. calories of heat evolved (show method of calculation) ______

h. water equivalent of the calorimeter ______

i. calories of heat absorbed in this reaction by the calorimeter ______

j. total number of calories of heat evolved (steps g and i) ______

k. ionic equation for the reaction between hydrochloric acid and sodium hydroxide

l. number of moles of water formed in the reaction using the quantities (in II a,b.) ______

m. number of calories of heat evolved per mole of water formed ______

III. The reaction between calcium and water.

a. weight of calcium ______

b. weight of water based on measured volume and accepted density ______

c. weight of reaction mixture ______

d. initial temperature of the water to nearest 0.2°C ______

e. final temperature of the mixture to nearest 0.2°C ______

f. difference in temperature between reactants and products to the nearest 0.2°C ______

g. calories of heat evolved (show method of calculation) ______

h. water equivalent of the calorimeter ______

i. calories of heat absorbed by the calorimeter in the reaction ______

j. total number of calories evolved for the reaction (sum of steps g and i) ______

k. ionic equation for the reaction of calcium with water ______

NAME______________________________SECTION____________DATE______________

l. number of moles of calcium used in the reaction _____

m. number of calories of heat evolved per mole of calcium (show method of calculation) _____

THOUGHT

1. Write net ionic equations for

   a. the reaction between hydrochloric acid and sodium hydroxide.

   b. the reaction between calcium and water (assume that the solution containing the calcium hydroxide is dilute and that the calcium hydroxide has completely dissociated into its ions).

   c. the reaction between calcium and hydrochloric acid.

   d. Arrange equations (a) and (b) above so that algebraic addition yields (c).

   e. Add your experimentally determined values for the molar heats of reaction as indicated by the equations above. This summation is the heat of reaction for calcium and hydrochloric acid, reaction (c).

NAME________________________ SECTION____________ DATE____________

Experiment 16
CHEMICAL THERMODYNAMICS

PRELIMINARY QUESTIONS

1. a) Using the following values for the solubility of lead chloride, calculate how many moles of lead chloride are in solution at these three temperatures.

   0°C = 0.673g/100 ml sol.
   20°C = 0.99g/100 ml sol.
   100°C = 3.34g/100 ml sol.

   b) What is the concentration of lead chloride in moles per liter in each of the solutions?

2. a) Write the chemical equation for solid lead chloride in equilibrium with its ions.

   b) Write the corresponding equilibrium-constant expression.

   c) What specific name is given to the equilibrium-constant expression for relatively insoluble salts?

d) Explain why only the ion product is used in this expression.

3. Using the values from question 1, calculate the solubility products for lead chloride at 0°, 20°, and 100°C.

Experiment 16
CHEMICAL THERMODYNAMICS

IDEAS

What causes chemical reactions? Why do different chemical substances have an affinity for each other? The attempt to find a satisfactory solution to this problem spurred many investigations and much thought. The insight that energy changes in chemical reactions can provide an important clue to this problem of chemical affinity was a triumph of late nineteenth century science. W. Nernst wrote*: "Certainly the explanation of the law of constant and multiple proportions and the marvelous systematics . . . are tremendous achievements of atomistics purely in the field of chemistry; . . . but these have hardly any relation to the manner in which two atoms combine in a compound, with the magnitude of the forces that enter in, and with the change of energy thus determined."

Chemical thermodynamics is the field of knowledge resulting from convergence of energy studies in physics related particularly to heat energy and studies of the heat of chemical reactions, thermochemistry. In their developmental stages, these appeared to be parallel but unrelated; the weaving together of these concepts into the coherent, systematic fabric of ideas called chemical thermodynamics was a great feat. The repercussions both within and outside of the field of chemistry were enormous.**

To provide a conceptual understanding of some important ideas in chemical thermodynamics emphasized in this experiment, the development of energy concepts in physics leading to thermodynamics, and the parallel progress in chemistry, will be briefly outlined.

The outline begins with the formulation of the principle of conservation of energy in the 1840's. By then, the concept of mechanical energy (ergon means "work" in Greek) and the equivalent concept of work had been defined after centuries of groping. However, it was not clearly understood that there are forms of energy other than mechanical energy. When the question of what heat is was finally settled in favor of the fruitful hypothesis that heat is a form of energy it led to the realization by some scientists that many forms of energy exist in nature. For instance, the German physician R. J. Mayer wrote: "In numberless cases we see motion cease without having caused another motion or the lifting of a weight; but an energy once in existence cannot be annihilated, it can only change its form; and the question therefore arises, what other form is energy, which we have become acquainted with as potential energy of kinetic energy capable of assuming."

---

* W. Nernst, "Thermodynamic Calculation of Chemical Affinities," in E. Farber, *Milestones of Modern Chemistry*, Basic Books, Inc., New York, 1966, p. 187.

** An excellent discussion of this is included in the Epilogue of L. K. Nash, *Elements of Chemical Thermodynamics*, Addison-Wesley Publishing Co., Reading, Mass., 1962. See also Henry A. Bent, *The Second Law*, Oxford University Press, New York, 1965.

In answer to his question, Mayer proposed that heat, chemical energy, and the energy of living and astronomical processes represent different forms of energy. The principle of conservation of energy is stated in the quotation from Mayer. This is one of the first statements of this principle, which, in one of its forms, later became known as the first law of thermodynamics. Simultaneously with Mayer, J. P. Joule, an English scientist, performed many experiments to measure the quantity of heat produced by work. He found that this ratio of heat to work was constant within the limits of experimental error, and he calculated this value (called the mechanical equivalent of heat) as 4.15 joules/calorie.* Realizing that the work done to produce different kinds of energy can be completely converted to heat, Joule made the induction, similar to Mayer's, that "the grand agents of nature are, by the Creator's fiat, **indestructible**, and that whenever mechanical force is expended, and exact equivalent of heat is always obtained." This principle of conservation of energy is a generalization based on countless observations.

Note also that the converse of Joule's statement is not valid; heat cannot be completely converted to work, which is one way of stating the principle that became known as the second law of thermodynamics.

The articulation of the second law was the culmination of even earlier studies in rather different channels on the relation between heat and work. N. S. Carnot, a French engineer, made an enormously important generalization in 1824, based on his study of steam engines, where heat is absorbed from a high-temperature reservoir. He affirmed that only a fraction of the available heat (which he at first considered to be caloric fluid) could be converted to work, and this fraction depends on the difference in temperature between the hottest and the coldest part of the engine. This statement provided the foundation of the field of thermodynamics (which means the motion of heat), and Carnot was one of the first to study the manner in which heat is converted to work. Ten years later, B. Clapeyron improved on Carnot's contributions by pointing out that the work done by the piston in the cylinder of the steam engine can be determined by plotting the pressure-volume curve and calculating the area under this curve.

The work of Carnot and Clapeyron, and the principle of conservation of energy provided in the 1850's the foundation of a new insight, the second law of thermodynamics. In Carnot's engine, only a fraction of the available heat can be converted to work. Although the system** may have heat, this heat cannot be used to perform work if its temperature is the same as that of its surroundings. This is not contrary to conservation of energy, for the energy of the system and its surroundings remains constant.

In the conversion of heat to work, only some of the available heat can be utilized. Some heat energy is lost to the surroundings, or we might say that it is wasted. Other energy conversion processes involve the incidental production of heat rather than its use. For instance, the operation of a motor to run a centrifuge produces some heat because of friction in the course of the conversion of mechanical energy to work. This heat tax must always be paid when mechanical energy is converted to work. As a result of friction, wrote W. Thomson (Lord Kelvin), "there is dissipation of mechanical energy, and full restoration to its primitive condition is impossible." In other words, some heat is always dissipated in the sense that it is no longer available to do work. This heat accumulates due to the incomplete conversion of heat to work, and also to

* See Appendix A for the present accepted value.

** The term "system" is one of several key words in thermodynamics. Other terms include "reversible" and "irreversible." A system is a region with definite, defined boundaries that isolate it from its environment. A container of gas, or the piston and cylinder of an engine, can be considered a system. If the volume and pressure are known, the temperature can be calculated. These variables are called the functions of the state of the system, and their values are independent of the way or path by which the state of the system is attained. A *reversible change* is one in which the system and its environment may be completely returned to their original conditions. This represents an idealization for observable systems. *Irreversible change* represents inexorable change.

the production of heat during other energy transformation processes. Consequently, although the total energy content of the universe is constant, the heat fraction of this total energy is always increasing.

Implicit in the preceding paragraph is the novel thought of irreversible, evolutionary processes in the universe. The universe itself as a system is in a state of irrevocable change. Because these concepts were the fruits of the studies of heat motion or transfer, Lord Kelvin named this field thermodynamics, and the statement that heat is dissipated or incompletely utilized in all transformations of heat to work is called the second law of thermodynamics. Alternative statements of the second law are that heat flows spontaneously from hotter to cooler bodies, or that work is required to transfer heat from a cooler to a warmer body.

R. Clausius, a German scientist, in 1865 introduced the term "entropy" to express the second law quantitatively. If Q is the heat absorbed or liberated, and T is the temperature, then, according to Carnot, for a reversible, cyclic process in a closed system

$$\frac{Q_a - Q_e}{Q_a} = \frac{T_a - T_e}{T_a} \quad \text{or} \quad \frac{Q_a}{T_a} = \frac{Q_e}{T_e}$$

Q_a is the heat absorbed at the higher temperature T_a; Q_e is the heat evolved at the lower temperature T_e. Clausius attached the symbol S to this factor Q/T and named it entropy. He wrote "I propose to designate S by the Greek word (which means) the transformation, the entropy of the body."* For a reversible process in an isolated system, the entropy, the transformation content, undergoes no change; but on the cosmic scale in which irreversible processes occur, "the entropy of the universe strives towards a maximum." In other words, physical processes have a spontaneous tendency to transform themselves into a state where no further change is possible. This is the state of maximum entropy.

The concept of entropy was originally applied to systems on the macroscopic scale, for example, heat engines. When the kinetic molecular theory of heat was accepted, the concept of entropy was reinterpreted in relation to the microworld in terms of statistical probabilities. For instance, consider a 22.4-liter container of an ideal gas at standard temperature and pressure: it contains 6×10^{23} molecules. This size sample allows the plotting of a good statistical distribution curve of velocity or energy states of the molecules versus the fraction of the total number. The most probable distribution of the velocities of the gas particles is the most random, or the distribution corresponding to maximum entropy—just as the most probable distribution of the cards in a deck is an unpatterned or random rather than an orderly arrangement. The concept of entropy is therefore extended to mean the tendency of processes to go in the most probable direction, to achieve the most probable statistical distribution.

The following equations are useful to summarize the first and second law of thermodynamics:

$$\text{First law:} \quad \Delta E = q - w$$

E is the internal energy of the system, which is a function of the state of the system; q is heat added to the system; and w is the work done by the system.

$$\text{Second law:} \quad \Delta S = \frac{q_{rev}}{T}$$

* R. Clausius, "Concerning . . . the Mechanical Heat Theory", in E. Farber *Milestones of Modern Chemistry*, Basic Books, Inc., New York, 1966, p. 155.

In this equation q_{rev} is the heat evolved or absorbed in a reversible change.

These developments in thermodynamics were completely isolated from chemistry until about 1869. In the meantime, there were concurrent advances in thermochemistry. For instance, many heats of reaction were determined, Hess's law was formulated (1840), and the French chemist H. Berthelot coined the terms "exothermic" and "endothermic." Berthelot considered that he had solved the problem of what causes chemical reactions when he proposed that all chemical reactions go in the direction of the maximum liberation of heat. Although it seemed to be true for many chemical reactions, this statement provoked controversy. It was suggested that the existence of chemical equilibrium offered a tangible refutation of that hypothesis. One side of a chemical equilibrium must be exothermic and the other endothermic; according to the hypothesis of Berthelot, only the exothermic reaction should proceed. Some reactions however, do proceed, even though heat is absorbed; they are endothermic rather than exothermic.

The application of thermodynamics to problems of what causes chemical reactions was a brilliantly successful union accomplished in the late nineteenth century by the American J. W. Gibbs and the Dutch scientist J. H. van't Hoff and others. Gibbs applied the concept of entropy to chemistry, basing his ideas on the work of Clausius. According to van't Hoff, a measure of chemical affinity is the maximum external work obtained from isothermal (at constant temperature), reversible chemical reactions. H. L. von Helmholtz had called this maximum work free energy, or available energy. In the early twentieth century the meaning of the term "free energy" was narrowed by Gilbert Lewis to useful work; and the symbols F or G (to honor Gibbs) are used. The concept of free energy thus became the criterion for evaluating chemical reactions.*

The fruition of these ideas on free energy and entropy in chemistry, and their relationship, was ultimately expressed in the elegantly simple expression $G = H - TS$, where G is the free energy or available energy, S the entropy, T the absolute temperature, and H the enthalpy. Since the study of chemical reactions involves the investigation of change, the equation becomes useful if it is expressed as $\Delta G = \Delta H - T\Delta S$ at constant temperature. ΔG is the difference between the free energies of the products and the free energies of the reactants. ΔH is the difference between the enthalpies of the products and the enthalpies of the reactants. ΔS is the difference between the entropies of the products and the entropies of the reactants.

INVESTIGATION

Purpose: To determine the changes in both standard enthalpy, $\Delta H°$, and standard entropy, $\Delta S°$, from the changes in solubility product with temperature.

Equipment:

- 1 250-ml beaker
- 3 50-ml beakers
- 1 110°C thermometer
- 3 watch glasses
- 1 glass marking pen
- 1 tripod or ring stand with ring clamp
- 1 hot plate

Chemicals: lead chloride, $PbCl_2$, solid

Procedure:

The outline of the procedure is simply to find out how much lead chloride is dissolved in a specific volume of solution at a known temperature by evaporation of the solution and weighing the residue.

---

* Suggested readings: B. Mahan, *University Chemistry*, Addison-Wesley Publishing Co., Reading, Mass., 1966. L. K. Nash, *Elements of Chemical Thermodynamics,* Addison-Wesley Publishing Co., Reading, Mass., 1962.

Clean and dry thoroughly three 50-ml beakers and three watch glasses. Calibrate ıe beakers by measuring 50 ml of water in a graduated cylinder, transferring he water to the beaker, and marking the 50 ml level with a glass marking pen-il. Empty the water from the beakers, dry them thoroughly, and number each equentially. Number the 3 watch glasses on their concave surfaces to correspond ith the numbers on the beakers. Weigh the marked beakers with their watch lasses to the nearest centigram.

Saturated solutions of lead chloride are prepared by adding about 5 grams f the salt to 400 ml of water in a 250-ml beaker. Heat the mixture to about 0°C on a hot plate and then while stirring rapidly, allow the solution to cool lowly to about 50°C.

You will notice that, although lead chloride is a dense salt, most of which ettles to the bottom of the beaker, a light scum persists on the surface. To void pouring this undissolved solute into the weighed beakers:

1) stir the solution to form a whirlpool to pull as much of the lead chloride as possible below the surface of the solution;

2) allow the solution to stand 2-3 minutes without stirring so that the precipitate can settle to the bottom of the beaker;

3) carefully skim the small amount of remaining surface scum off with a piece of filter paper.

4) when pouring the solution from the 250-ml beaker into the 50-ml beaker, place the thermometer or stirring rod across the top so that it rests on the lip of the beaker (see figure 2). When you pour the solution through the lip, the liquid will trail down the rod or thermometer. The direction of the solution can be aimed precisely into the beaker, and the scum on the surface will be held back by the rod.

When the solution has reached 50°C, pour off as rapidly as possible 50 ml nto the weighed beaker (#1). Cover with the appropriate watch glass, and vaporate the solution slowly on the hot plate until there is splattering. Re-ove the last bit of water from the residue by placing the beaker and watch lass in an oven set at 110°C-120°C for several hours. Observe the evaporation n the oven frequently to see if a "lid" of lead chloride has formed over the olution preventing evaporation. If this "lid" has formed, use tongs and gently ove the beaker back and forth to dislodge the lead chloride from the beaker alls. It will then fall to the bottom and evaporation can continue.

While the solution is evaporating, proceed in a similar way to remove 50 ml f solution at 40°C and 30°C for the remaining 2 samples.

Record all data on data sheets.

Discussion of Calculations

To help you understand how you can develop the values for free energy, nthalpy, and entropy from the experimental data, the table below defines the ymbols used in the quantitative development which follows.

Symbol	Definition	Units
G	Free energy	kJ/mole
G°	Standard* free energy	kJ/mole
S	Entropy	J/(mole)(deg)
S°	Standard* entropy	J/(mole)(deg)
H	Enthalpy	kJ/mole
H°	Standard* enthalpy	kJ/mole
K_{eq}	Equilibrium constant	
R	Gas constant	8.31 kJ/(mole)(deg)
ln	Natural log, base e	
$\log_{10}$	Log to the base 10	
T	Absolute temperature	°K

$$\Delta G^\circ = \Delta H^\circ - T \Delta S^\circ \qquad (1)$$

A general chemical equation may be written as:

$$aA + bB \rightarrow cC + dD$$

If this is written in a mass-action expression

$$\frac{[C]^c [D]^d}{[A]^a [B]^b}$$

and represents a system not at equilibrium, then

$$\Delta G = \Delta G^\circ + RT \ln \frac{[C]^c [D]^d}{[A]^a [B]^b} \qquad (2)$$

If equilibrium exists, then $\Delta G = 0$ and

$$\Delta G^\circ = -RT \ln \frac{[C]^c [D]^d}{[A]^a [B]^b} \qquad (3)$$

$$\Delta G^\circ = -RT \ln K_{eq} \qquad (4)$$

* Standard states of a substance are defined as the stable state of that substance at 298°K and one atmosphere pressure. The superscript '°' is used to indicate that the value for the thermodynamic function applies to substances in their standard states.

Combining Eq. (1) and (4)

$$-RT \ln K_{eq} = \Delta H° - T\,\Delta S° \quad (5)$$

Multiply through by -1:

$$RT \ln K_{eq} = -\,\Delta H° + T\,\Delta S° \quad (6)$$

$$\ln K_{eq} = -\frac{\Delta H°}{RT} + \frac{T\,\Delta S°}{RT} \quad (7)$$

$$\ln = 2.303 \log_{10}$$

$$\log_{10} K_{eq} = -\frac{\Delta H°}{2.303R}\,\frac{1}{T} + \frac{\Delta S°}{2.303R} \quad (8)$$

Equation (8) resembles the algebraic equation for a straight line

$$y = mx + b$$

in which m is the slope of the line and b is the y intercept. Inspection of Eq. (8) shows that if the values for $\log_{10} K_{eq}$ are plotted against the values for 1/T, the slope m is $\Delta H°/2.303R$ and the y intercept is $\Delta S°/2.303R$.*

---

* The experimental data provide values for temperature ranges between 0°C and 100°C. Therefore, all interpretations of ΔH and ΔS are based upon only this narrow range of data. Since Eq. (8) resembles the algebraic equation for a straight line, $\Delta S°$ is sometimes computed by extending the curve until the value for 1/T = 0 and then calculating the value of $\Delta S°$ using the y intercept. This method in this experiment yields a value for $\Delta S°$ that is negative. So, although it would appear that $\Delta S°$ can be found graphically by extrapolating to the y axis, there is some question as to the validity of method. At the y intercept, 1/T implies a value of infinite temperature for T, which has no physical meaning. It also implies that 1/T vs the log of equilibrium constant is a straight line for all values of T. Therefore, to calculate $\Delta S°$, use equation 8.

In this system several processes must be considered.

1. The entropy change resulting when the ions of the crystal leave their positions and become more randomly distributed in the solution. The entropy of the solute would increase and ΔS is positive.

2. The entropy change in the solvent, resulting from the orientation of the polar water molecules in a pattern around each ion. Thus the entropy of the solvent, ΔS, decreases and is negative.

NAME____________________ SECTION__________ DATE__________

CHEMICAL THERMODYNAMICS

DATA

Sample	1	2	3
Temperature °C			
Weight of beaker and watch glass + $PbCl_2$			
Weight of beaker and watch glass			
Weight of $PbCl_2$			

Calculations

1. Complete the following table for your three experimental values and the three accepted values from the table of solubilities. (Appendix F)

ample	Temp (°C)	Vol. Sol.	Weight $PbCl_2$	Moles $PbCl_2$	Conc. $PbCl_2$ moles/liter	Conc. Pb^{2+} moles/liter	Conc. Cl^- moles/liter	K_{sp}* $PbCl_2$	$\log_{10} K_{sp}$	T	$\frac{1}{T}$
rimental											
1											
2											
3											
pted											
1	0										
2	20°										
3	100°**										

e equilibrium expression for a sparingly soluble salt like lead chloride can be written as $K_{eq} = \frac{[Pb^{2+}_{(aq)}][Cl^-_{(aq)}]}{[PbCl_{2(s)}]}$

ad chloride is a solid and its concentration (moles/liter) is therefore constant. Hence a new expression can be ived defining the solubility product constant, K_{sp}.

$$(K_{eq})[PbCl_{2(s)}] = [Pb^{2+}_{(aq)}][Cl^-_{(aq)}]^2 = K_{sp}$$

cause lead chloride is quite soluble at 100°C, there may be some question about the use of this data to compute olubility product. In this experiment it is assumed that activity and molar concentrations are nearly the same.

2. a. Plot the values for $\log_{10} K_{sp}$ on the y-axis and the values for 1/T on the x-axis. With a ruler draw the best possible straight line which connects or passes near as many points as possible.

   b. Calculate the slope of this line and ΔH°.

3. Using equation (4), calculate ΔG° for 298°K.

4. Using equation (8), calculate ΔS° for 298°K.

THOUGHT

1. a. Considering your value for ΔH, is the solution of lead chloride an endothermic or exothermic process?

   b. Is ΔS positive or negative (i.e., increasing or decreasing)?

   c. Is ΔG positive or negative (i.e., not spontaneous or spontaneous)?

2. a. Does the solution of $PbCl_2$ take place spontaneously at 25°C? How do you know?

NAME______________________ SECTION__________ DATE__________

b. Could you have predicted the spontaneity of the reaction by knowing only ΔH?
or only ΔS?
or only ΔG?

3. Is the ΔH for this reaction the same as the ΔH of formation for lead chloride from its elements? Explain.

4. In the dissolving of $PbCl_2$, describe two simultaneous processes which affect the experimentally determined entropy of the system.

NAME______________________________SECTION__________DATE__________

INTRODUCTION TO SPECTROPHOTOMETRIC ANALYSIS

(EXPERIMENTS 17 to 19)

PRELIMINARY QUESTIONS

1. Define the following terms:*

   a) wavelength

   b) frequency

   c) visible spectrum

   d) complementary color

   e) monochromatic light

   f) absorbance

   g) transmittance

   h) spectrophotometer

---

* See Introduction to Spectrophotometric Analysis.

2. What determines the color of a solution?

3. What determines the intensity of the color of a particular solution?

4. Percent transmittance has the following mathematical relationship to absorbance:

$$A = \log \frac{100}{\%T}$$

a) What are the absorbances for the following percent transmittances?

i. 100 percent

ii. 1 percent

iii. 80 percent

b) What is the percent transmittance if the absorbance is 0.3?

INTRODUCTION TO SPECTROPHOTOMETRIC ANALYSIS

IDEAS

The roots of the ideas which converged to produce the spectrophotometer can be traced to the birth of modern science in the seventeenth century when Newton devised a series of brilliant experiments to investigate the nature of light and of color. The emergence of the spectrophotometer as a commonly used tool of analytical chemistry did not occur until the 1940's, almost 300 years later. During that interval of about three centuries, the accumulating ideas that led to the idea of the spectrophotometer and its applications penetrated the very marrow of the physical sciences, because they were based on attempts to answer such fundamental questions as the nature of light, the nature of matter, and the nature of their interaction.

Some of the highlights of this long and complex story include: (1) the recognition that white light is composed of many colors; (2) the development of successively better theories of light and on the interaction of light and matter; (3) the use of the phototube to detect light intensity with precision; (4) the use of empirical methods of colorimetry in chemistry; and (5) the study of light absorption in solutions and other transparent media.

1. That white light is composed of the colors of the rainbow was recognized by Newton in the seventeenth century. Through his beautifully ingenious prism experiments, Newton demonstrated that white light passing through a prism, is broken up into a spectrum of colors. He wrote with enthusiasm:

> But the most surprising and wonderful composition was that of Whiteness. There is no one sort of Rays which alone can exhibit this. 'Tis ever compounded, and to its composition are requisite all the aforesaid primary Colours mixed in a due proportion. I have often with Admiration beheld, that all of the Colours of the Prism being made to converge . . . reproduced light, intirely and perfectly white . . . .*

2. The realization that white light is made up of colors intensified the problem on the nature of light which Newton and Huygens attempted to solve when both offered competing theories in the seventeenth century.** These were the first theories of light of the modern scientific era, and they were successively modified over a period of about 300 years. Newton

* "A Letter of Mr. Isaac Newton . . ., " in *Issac Newton's Papers and Letters . . .*, I. B. Cohen, ed., Harvard University Press, Cambridge, Mass., 1958, p. 55.

** Sparberg, E., Misinterpretation of Theories of Light, *American Journal of Physics*, May, 1966, pp. 377-389.

proposed that light may be made up of streams of particles or corpuscles, and one of his great problems was to account for their passage through transparent media such as water or glass, which he did by assuming that they were carried through the rare fluid aether that surround the atoms of all matter. Another problem was how to explain the color of a transparent medium in terms of his theory. His explanations, which were very complex,* were based upon the production of vibrations of various sizes within the aether permeating these bodies. Huygens had proposed a kind of wave theory of light; by the middle of the nineteenth century a drastic revision of a wave theory of light was accepted, which was transformed in the 1860's into the modern electromagnetic wave theory.

3. Einstein's interpretation of puzzling phenomena on the production of electrical effects by light and further experimental investigations of Einstein's predictions by Robert Millikin led to the construction and use of sensitive phototubes. Although the wave nature of light was firmly accepted at this time, experimental evidence had led Einstein to propose in 1905 that light under certain experimental conditions exhibits the characteristics of particles. Planck had earlier formulated the quantum theory, in which he made the assumption that energy is discrete or particle-like in nature in the subatomic microworld. Allowing light to bombard the surface of various metals could cause the expulsion of electrons from these metals, and Einstein suggested that this photoelectric effect might best be interpreted according to the quantum model rather than the wave model. He suggested that light is composed of streams of quanta, or photons as they were later called. Einstein's theory appears to have come full circle round to Newton's after centuries of revisions, but in a sense there are only certain philosophical similarities, for the photons of contemporary science have properties quite different from the corpuscles of Newton. For instance, according to the modern interpretation, a blue solution absorbs photons of all frequencies of light with the exception of blue, which is transmitted through the medium. In terms of the present model of the atom, which is the result of a continuous series of revisions of the Bohr atom, the electron clouds surrounding the nuclei of the solute particles can absorb or eject energy packets of specified sizes only. The blue photons do not meet the energy requirements, and they may travel freely through the solution to meet their doom upon striking the cathode of a blue-sensitive phototube. But their annihilation is not in vain, for their energy has been fully utilized to eject electrons with varying amounts of kinetic energy from the cathode. As discussed in the next section, the intensity of the light, which is the number of photons striking the cathode per second, determines the number of electrons expelled per second. The latter constitutes the electric current produced.

4. Many years before the theoretical implications of the color of solutions were adequately understood, colorimetric techniques were used empirically for quantitative work in chemistry. For instance, the red iron (III) thiocyanate complex ion and the tetraammine copper (II) complex ion were used for analytical purposes by the middle of the nineteenth century, and visual methods were used to compare unknown solutions with those of known concentrations. The analytical detection of ammonia by the formation of a yellow precipitate with a basic solution of potassium mercury (II) iodide (Nessler's method) was used from about 1852. Even today, Nessler tubes, which are about the oldest kind of apparatus used for visual color comparisons, are not uncommon as tools of quantitative analysis.

5. The spectrophotometer, which was used widely by about 1941, represented a great improvement over empirical, visual methods, as the phototube in the instrument can measure with great precision the intensity of light emerging from a solution. The study of the light

* "Newton's Second Paper on Colour and on Light," *Isaac Newton's Papers and Letters* . . . , Ibid., pp. 192-193.

absorption properties of a medium is therefore yet another one of the important preliminary areas of investigation contributing to the development and use of the instrument. P. Bouguer and J. Lambert discovered in the eighteenth century that a medium absorbs only a fraction of the incident light, and that the fraction absorbed is proportional to the thickness of the medium. A. Beer in 1852 found that the absorption depends also upon the concentration of the solute. The quantitative relationship $A = abC$, described in the following discussion, is usually known as Beer's law, although it incorporates not only Beer's findings but those of Bouguer and Lambert.

The name "spectrophotometer" gives a good clue on the nature of this instrument: "spectro" related to color, "photo" meaning light, and "meter" signifying measurement. Light of one color, called monochromatic, is allowed to travel through a transparent medium, such as a solution, and the intensity (I) or power (P) of the light that emerges is measured. During its journey through the solution, some light may be absorbed. In addition, some light is also reflected from the surfaces. The intensity of the emerging light depends mostly upon the amounts of light absorbed by the solution and reflected from its surface.

The question arises of why transparent materials such as solutions may differentially absorb various wavelengths of visible light. The wave length of light can be defined as the distance between identical positions on adjacent waves. Since the different wavelengths have different energy values, the process involves the absorption of energy. According to the quantum theory of light, a beam of white light, which is composed of all the colors of the visible spectrum, consists of streams of photons. The energy E of each photon, by the Planck equation $E = h\nu$, is directly proportional to the frequency ν of the light, where h is the proportionality constant called Planck's constant. Frequency is defined as the number of waves which pass by a point in a unit of time.

The relation between wavelength and frequency of light is given by the equation $c = \lambda\nu$, where c is the speed of light, λ the wavelength, and ν the frequency. Recall that various colors are characterized by different wavelengths of light. The visible part of the electromagnetic spectrum ranges from a wavelength of about 350 nanometer for violet light to about 750 nm for red.

The selective absorption of light photons by solute particles indicates that the latter absorb energy quanta of only certain sizes. The sizes that are likely to be absorbed are those which correspond precisely to the most probable transition of the atom, ion, or molecule to a higher energy level, or the excited state. In the process of absorbing photons of visible light to produce the excited state, outer electrons may have been pumped to a higher energy level, and vibrational and rotational levels may have also changed. The excited atoms spontaneously return to the ground state, giving off their energy, usually in the form of heat. Because the outer electrons are responsive to visible light, the manner in which an atom has combined chemically greatly affects what light photons are absorbed, and hence the color of the compound. For instance, iron (II) thiocyanate is green, while iron (III) thiocyanate is red.

The color of a particular solution may be produced by the kind and number of solute particles present in a colorless solvent. These particles may absorb some of the wavelengths in a beam of white light passing through a solution, and transmit others. The color of the solution is the band of wavelengths that has been transmitted, which is said to be the complement of the absorbed light because the reunion of the absorbed and transmitted wavelengths would produce the original white light. For instance, copper sulfate pentahydrate appears blue when white light is allowed to pass through an aqueous solution because all wavelengths of light are absorbed by the copper ions except a band of blue, which is transmitted through the solution. The number of copper ions per unit volume determines the intensity of the blue color; the smaller their number per unit volume, the fewer there are available for the absorption of wavelengths other than blue. Therefore, more light of all wavelengths is transmitted, and consequently the blue color is diluted and appears lighter. A deep blue color, by this reasoning, indicates larger numbers of copper ions, solute particles, in a given volume.

Because the kinds of particles in solution and their number profoundly influences the color of the solution, and its intensity as discussed above, quantitative analysis of colored solutions under some conditions may be successfully accomplished by the use of the spectrophotometer. Many inorganic ions, especially those of the transition elements, form colored complexes in solution which can be quantitatively analyzed with a high degree of accuracy by use of the spectrophotometer. In certain cases where trace ions are present in a concentration as low as 10^{-5} percent, this instrument provides a valuable tool for use in quantitative analysis. Its application hinges on the relationship between the intensities of the incident and emergent beams, known as Beer's law: in dilute solutions, the absorbance (A) of the solution is directly proportional to the concentration (C) of the absorbing particles when the length of the light path is constant. Beer's law may be expressed: $A = abC$, where a is a constant called the absorptivity, b is the length of the light path in centimeters, and C is the concentration expressed in grams per liter.

It is necessary at this point to clarify the equivalent terms of light intensity and power or radiant power; the intensity (I) or power (P) of a beam of light is the radiant energy (E) impinging on a unit area per unit time (t). Power is defined generally as energy per unit time: $P = E/t$. Since each photon is equivalent to a specific amount of energy, the intensity of light is also proportional to the number of photons bombarding the surface per unit time. The technique to measure the absorbance of a solution containing colored solute particles is as follows: (1) The absorbance of the pure solvent which may also absorb light energy is measured first. This absorbance is a reference. (2) The absorbance of the solution is then measured. (3) Any difference between the absorbance of the pure solvent and the solution can be attributed to the presence of the solute species.*

As mentioned above, Beer's law can only be used for dilute solutions, and there are other theoretical and experimental limitations.** One of the limitations is the necessity to use monochromatic light for the incident beam. If a polychromatic beam is used, the absorbing species may selectively absorb different wavelengths. Therefore, the direct proportionality between absorbance and concentration is no longer valid.

The use of monochromatic light has two additional advantages. Interference caused by the presence of other species is minimized if the latter absorb at different wavelengths. Second, maximum sensitivity results from the use of the wavelengths of maximum absorption for the species under analysis, because any change in concentration is easily detected. For instance, the most intense blue Ni-EDTA solution is selected and bombarded with light of various wavelengths. The wavelength most strongly absorbed, which will be in the red range, is selected as the most suitable monochromatic beam. The less intense blue solutions will absorb fewer red photons, and so the red-sensitive detector will indicate a greater light intensity.

The spectrophotometer consists essentially of (1) a light source; (2) a diffraction grating or prism; (3) a monochromator which isolates a band of wavelengths from the spectrum produced by a prism or diffraction grating; (4) transparent sample containers for the liquids to be analyzed; and (5) a phototube to detect the radiation after transmission through the liquid. The cathode of the phototube has a cesium-antimony surface which can expel outer electrons when bombarded by photons of sufficient energy. The particular coating is sensitive to particular

* Transmittance (T) has been defined as the fraction or percent of incident radiation transmitted by the solution. $P/P_o = T$; $A = \log(1/T) = \log(P_o/P)$, where P_o is the incident power or intensity.

** There are good discussions of this in such texts by D. Skoog and D. West, *Fundamentals of Analytical Chemistry*, Holt, Rinehart & Winston, New York, 1963; R. A. Day, Jr., and A. L. Underwood, *Quantitative Analysis*, Prentice-Hall, Inc., Englewood Cliffs, N. J., 1967; and G. W. Ewing, *Instrumental Methods of Chemical Analysis*, McGraw-Hill Book Co., Inc., New York, 1969.

frequencies; for example, the red-sensitive phototube responds to red light, which has a lower energy value than blue light. The number of electrons expelled, which is the current produced in the phototube, is directly proportional to the intensity of the light or the number of photons hitting the cathode per second.

The Spectronic 20 spectrophotometer and a schematic diagram of its internal mechanism are pictured in Figures 21 and 22. It produces and utilizes a spectral band width of 10 to 20 nm and its range is from 350 to 950 nm with the appropriate phototubes.

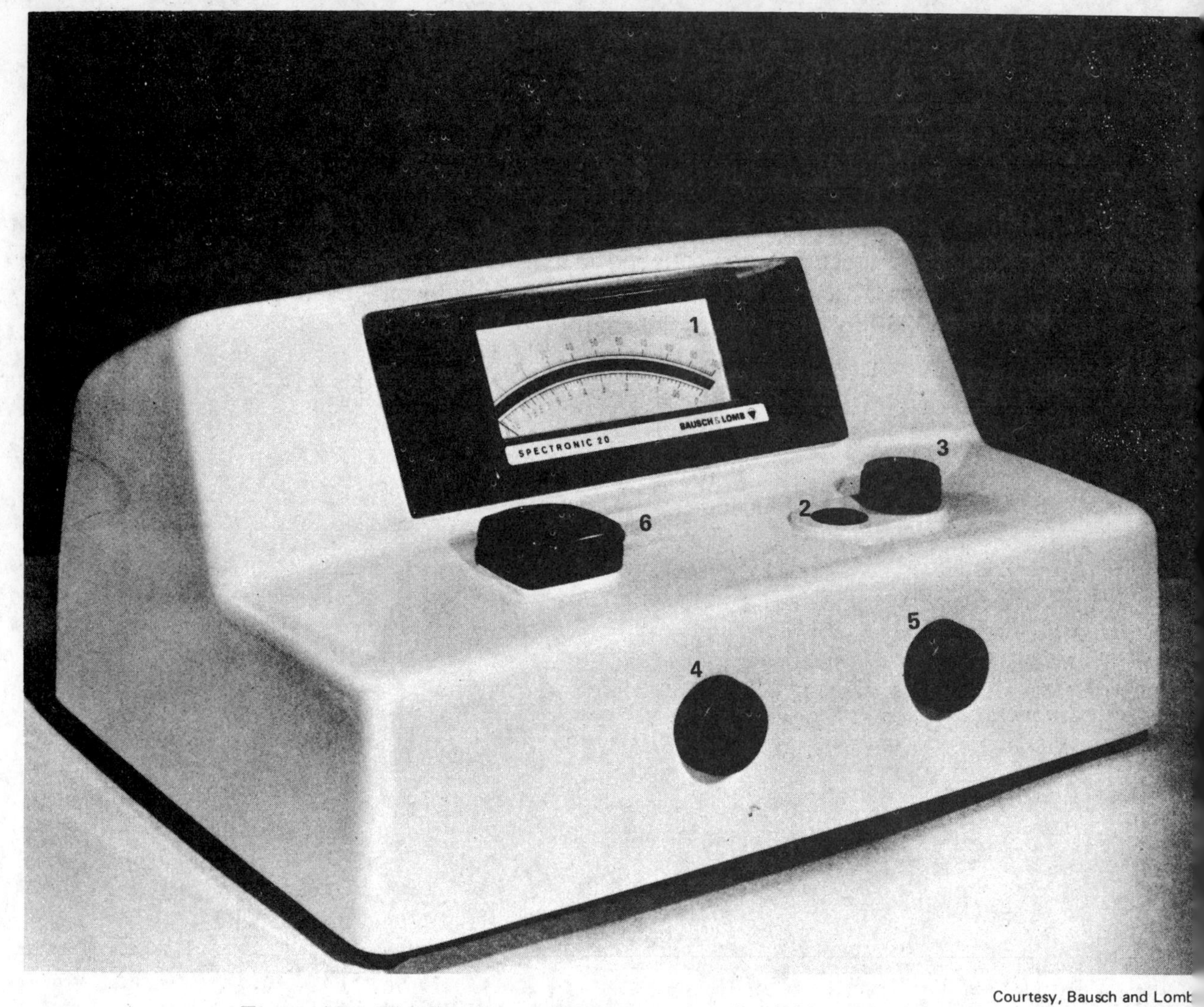

Courtesy, Bausch and Lomb

Figure 15. The Spectronic 20 Colorimeter/Spectrophotometer

1. Meter
2. Wavelength Scale
3. Wavelength Control
4. Light Control (on-off switch)
5. Power Switch & Zero Control
6. Sample Holder

LIGHT PATH AS VIEWED FROM A POINT DIRECTLY ABOVE THE INSTRUMENT

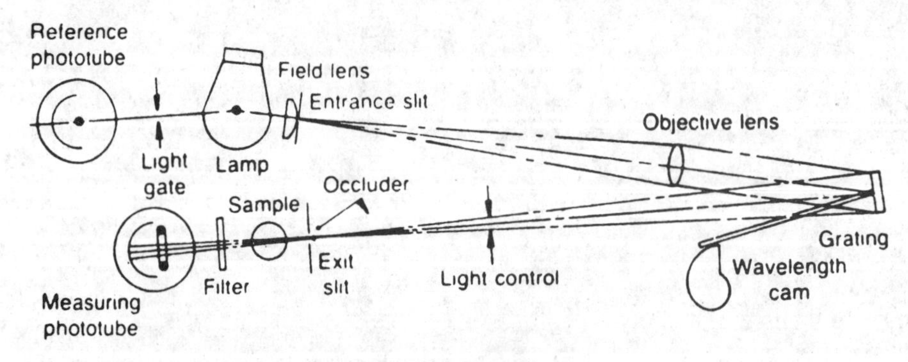

(Courtesy of Bausch & Lomb.)

Figure 16. The Spectronic 20 has three convenient control knobs, and the readings are made directly from the meter in either absorbance or transmittance.

NAME______________________________ SECTION____________ DATE____________

Experiment 17
SPECTROPHOTOMETRIC ANALYSIS OF AIR POLLUTANTS

PRELIMINARY QUESTIONS

1. What are the air pollutants which are being analyzed in this experiment?

2. Where would you expect to find heavy concentrations of those pollutants?

3. Define turbidity.

4. What are the products of the reaction between the air pollutants and potassium carbonate?

5. How is sulfur dioxide detected? Write balanced equations for the chemical reactions.

6. How is nitrogen dioxide detected?

7. Explain the function of the $Hg(SCN)_2$ and the iron(III) alum in the analysis of the chloride ion.

Experiment 17
SPECTROPHOTOMETRIC ANALYSIS OF AIR POLLUTANTS

IDEAS

The problems of air pollution are not new. In the late thirteenth century, a pollution problem was caused in London by smoke from the burning of coal. This was not ignored, and attempts to control pollution included the taxing of coal. By the seventeenth century, King Charles II of England requested that Sir John Evelyn, a noted scholar, prepare a report on the problem. It was entitled "On Dispelling of Smoke" (1661), and the title page quoted the Roman poet Lucretius of the first century B.C. "How easily the heavy potency and odor of the carbons sneak into the brain." Sir Evelyn wrote about the disastrous effects of London smoke, ". . . the inhabitants breathe nothing but an impure thick mist, accompanied with a fuliginous and filthy vapor, corrupting the lungs. Coughs and consumption rage more in this one city than in the whole world."

But the situation did not improve. Industrialization, urbanization, and the explosion of technology exacerbated the problem. Various serious air pollution crises occurred in this century. In London, in 1952, over 4000 deaths were attributed to a cold thick fog which prevented smoke containing sulfur dioxide, and other pollutants from rising into the upper atmosphere. A similar disaster had occurred in Donora, Pennsylvania in 1948, but in a less densely populated area.

The catastrophes above were caused mainly by sulfur dioxide mixed with smoke, soot, and other matter, which is produced by the burning of coal and oil. This mixture is called London type smog. Another type of smog common in Los Angeles is called photochemical smog; sunlight initiates reactions among pollutants emitted mostly by the automobile. These are mainly nitrogen oxides, ozones, hydrocarbons and other organic compounds.

In today's experiment, you will investigate whether sulfur dioxide, nitrogen dioxide, and chlorine are pollutants in your environment. You will be able to determine semi-quantitatively if these pollutants are present, and whether their concentration is small, moderate or large relative to the quantities found by other members of the class.

In recent years, the Clean Air Act (1963, amended 1977) has made it mandatory to reduce the amount of sulfur found in most oil and coal. The effect has been a perceptible reduction in the sulfur dioxide content of the atmosphere. An increasing shortage of petroleum in oil importing countries due to international political problems has focused attention on the use of coal. Most raw coal contains some sulfur necessitating its chemical removal before burning. This will profoundly affect the economics of energy.

Nitrogen oxides can be emitted by the automobile, since both oxygen and nitrogen of the atmosphere are burned in the high temperature cylinder. The source of chlorine pollutants is the many chlorine-containing agents used for bleaching and purification, e.g., bleach for clothes, cleansers, swimming pool bactericides, and chlorinated water. Their pervasive odor indicates their volatility.

Common air pollutants in industrial areas and areas of high traffic concentration may include chlorine and the oxides of nitrogen and sulfur. One method of detection and quantitative determination of these gases consists simply of exposing a glass fiber filter which has been impregnated with potassium carbonate to the atmosphere for 2-4 weeks. The sulfur oxides are converted to barium sulfate and its concentration is measured in terms of the turbidity of an aqueous suspension. Both nitrogen dioxide and chlorine are converted to colored compounds whose solution concentration can be determined by the amount of light of a specific wavelength which is absorbed. All the measurements are made by using a spectrophotometer.

Discussion of Some of the Chemical Reactions

1. Reaction of the pollutants with potassium carbonate.

The pollutants which are investigated in this experiment are all acidic anhydrides and can react with moisture to form acids. For example:

$$SO_2 + H_2O \rightarrow H_2SO_3$$

$$2SO_2 + O_2 \rightarrow 2SO_3$$

$$SO_3 + H_2O \rightarrow H_2SO_4$$

$$2NO_2 + H_2O \rightarrow HNO_2 + HNO_3$$

$$3NO_2 + H_2O \rightarrow NO + 2HNO_3$$

$$2NO + O_2 \rightarrow 2NO_2$$

$$Cl_2 + H_2O \rightarrow HClO + HCl$$

These acids will react with the carbonate ion to form a salt, carbon dioxide, and water. The anhydrides of these acids react to form only the salt and carbon dioxide.

Reactions with potassium carbonate may be represented by the following equations:

$$K_2CO_3 + SO_2 \rightarrow K_2SO_3 + CO_2$$

$$K_2CO_3 + SO_3 \rightarrow K_2SO_4 + CO_2$$

$$K_2CO_3 + 2HNO_2 \rightarrow 2KNO_2 + H_2CO_3$$

$$K_2CO_3 + 2HCl \rightarrow 2KCl + H_2CO_3$$

$$H_2CO_3 \rightarrow CO_2 + H_2O$$

During the exposure period it is assumed that most of the sulfur dioxide has been oxidized to sulfur trioxide, that the oxides of nitrogen have been converted to the nitrite ion, and that chlorine is present as the chloride ion.

2. Detection of the Products

When the glass fiber filter is ready for analysis, the soluble salts are extracted with boiling water. An especially sensitive test is used to detect each ion. The reactions for each follow.

a. the sulfate ion

$$SO_4^{2-} + Ba^{2+} \rightarrow BaSO_4$$

Barium sulfate is a white crystalline precipitate whose concentration can be determined spectrophotometrically by the turbidity of its suspensions. Sulfite ion (from SO_2) which has not oxidized to the sulfate ion during air exposure, is oxidized with hydrogen peroxide.

$$H_2O_2 + SO_3^{2-} \rightarrow SO_4^{2-} + H_2O$$

Since barium carbonate is also a white crystalline material which if present would appear indistinguishable from barium sulfate, it must be dissolved in hydrochloric acid.

$$CO_3^{2-} + 2H^+ \rightarrow H_2O + CO_2$$

This reaction is characteristic of that for salts of a weak acid in the presence of a strong acid.

b. the nitrite ion

The nitrite ion reacts with sulfanilimide to form a diazonium salt. This salt reacts with a primary amine, N(1-naphthyl) ethylene diamine dihydrochloride to form a highly colored azo dye.*

c. the chloride ion

The detection of the chloride ion depends upon the following reactions:

1. mercury (II) thiocyanate reacts with chloride ions to produce a weakly ionized salt, thus effectively removing the chloride ion from the solution and replacing it with the thiocyanate ion.

$$Hg(SCN)_2 + 2Cl^- \rightarrow HgCl_2 + 2SCN^-$$

---

*

$$H_2N{-}SO_2{-}C_6H_4{-}NH_2 + NaNO_2 + 2HCl \rightarrow H_2N{-}SO_2{-}C_6H_4{-}N_2^+Cl^- + NaCl + 2H_2O$$

Sulfanilimide — Diazonium salt

$$H_2N{-}SO_2{-}C_6H_4{-}N_2^+Cl^- + C_{10}H_7{-}NH{\cdot}HCl{-}CH_2{-}CH_2{-}NH_2\,HCl \xrightarrow[\text{acid}]{\text{mildly}} H_2N{-}SO_2{-}C_6H_4{-}N{=}N{-}C_{10}H_6{-}NH{\cdot}HCl{-}CH_2{-}CH_2{-}NH_2{\cdot}HCl$$

N(l-naphthyl) ethylene diamine dihydrochloride — Azo dye

2. Ferric ion, produced by the ionization of iron (III) alum, $FeNH_4(SO_4)_2\ 12H_2O$, reacts with thiocyanate ion, and produces a deep red solution.

$$Fe^{3+} + SCN^- \rightarrow FeSCN^{2+} \quad \text{(red solution)}$$

INVESTIGATION

Sulfur Dioxide Determination

Purpose: To determine whether sulfur dioxide is present in the atmosphere and its concentration.

Equipment:

47 mm glass fiber filter	1 1-liter volumetric flask
48 mm plastic petri dish	1 100-ml volumetric flask
50-ml beaker	5 15-cm test tubes
11-cm qualitative filter paper	3 1-ml transfer pipets
1 50-ml volumetric flask	2 50-ml burets
	Spectronic 20 instrument

Chemicals (Per Student):

Acetone	2 ml
Potassium carbonate, 30%, K_2CO_3	2 ml
Sodium sulfate, anhy, Na_2SO_4	2 g
Hydrochloric acid, HCl, 1.0 M	26 ml
Barium chloride, $BaCl_2 \cdot 2H_2O$	2.4 g
Hydrogen peroxide, 3%, H_2O_2	7.0 ml

Outline of Procedure

There are three parts to the analysis. The first part is the exposure of the potassium carbonate impregnated glass fiber filter to the atmosphere. The second part is the preparation of a series of solutions in which the sulfate concentration is known and upon which turbidity measurements can be made. A spectrophotometer is used to determine the turbidity of the solution. A plot on graph paper of the sulfate concentration vs absorbance gives a calibration curve. The third part is the preparation of the unknown solution for analysis, the measurement of the turbidity and the determination of the SO_2 concentration from the calibration curve of the SO_4^{2-}.

Procedure:

PART I. PREPARATION OF THE TEST DETECTION PLATE. *(Note: Avoid contamination of the plate. Use very clean tongs to handle the glass fiber filter.)*

Attach the glass fiber filter to the bottom half of a plastic petri dish with a few drops of acetone. Add 1 ml of 30% potassium carbonate solution and dry in an oven at 60°C. When dry, replace the cover and move the dish to the test area. During the 2-4 week exposure period the uncovered petri dish is fixed securely by tape in an inverted position so that particulate matter will not contaminate the plate and so that the atmosphere can circulate freely over the filter.

PART II. PREPARATION OF THE SOLUTIONS FOR THE CALIBRATION CURVE.*

a. Preparation of solution IIA

Prepare an aqueous solution of sodium sulfate by dissolving 0.740 g anhydrous sodium sulfate in a 1-liter volumetric flask. **Use only deionized water.**

b. Preparation of solution IIB

Dilute 10 ml of the solution prepared in (a) to 100 ml in a volumetric flask. What is the sulfate concentration in g/liter in this solution?

c. Using the diluted solution from (b) prepare a series of new solutions in which the sulfate concentration is increased by a constant factor as follows. Set up 5 test tubes, number them consecutively, and dilute according to the table using 1-ml pipets and burets.

test tube number	ml Na_2SO_4	ml 3% H_2O_2	1.0 M HCl ml	0.4 M $BaCl_2$ ml	ml H_2O	Total ml solution
(blank)	0	1	1	1	18	21
2	4	1	1	1	14	21
3	8	1	1	1	10	21
4	12	1	1	1	6	21
5	16	1	1	1	2	21

Shake each tube and let it stand 20 minutes.

Adjust the Spectronic 20 to read zero absorbance at 450 nm with the sample in test tube #1. This is the blank solution and deletes from the reading the contribution of all absorbing species which may be present in the solvent medium. Then measure the absorbances of the remaining solutions and complete the table on the data page.

PART III. PREPARATION OF THE TEST SAMPLE

a. Preparation of the test solution (IIIA)

When the exposure time of the test plate has ended, the filter is carefully removed from the dish with very clean tongs and spatula, placed in a 50-ml beaker containing 35 ml water and gently heated for 30 minutes. *Do not boil the solution.* The resulting solution is filtered through qualitative filter paper into a 50-ml graduated cylinder and the filter paper rinsed with two 5-ml portions of water which are added to the solution in the graduated cylinder. The volume is then brought to the 50 ml mark by the addition of deionized water. Prepare a blank solution in exactly the same way using a glass fiber filter, a few drops

* The preparation of the solutions and the measurements of the absorbances may be made in advance by laboratory personnel and distributed for student use.

of acetone and one ml of 30% potassium carbonate. Since these solutions are used not only for the SO_4^{2-} determination but also for the NO_2^- and Cl^-, *do not discard them until all the tests have been completed.*

b. Solution IIIB

Using a 10-ml graduated cylinder measure 7 ml of the solution in 'a' and put it into a clean dry container. Add 13 ml water and one ml 1.0 M HCl to it. Prepare the blank solution similarly from the blank solution prepared in 'a'.

c. Solution IIIC

Using a 10-ml graduated cylinder measure 7 ml of the solution prepared in IIIB and put it into a clean dry container. Dilute this solution with 11 ml water, and add one ml 0.4 M $BaCl_2$ solution and one ml H_2O_2. To adjust the acidity, also add 0.7 ml of the 1.0 M HCl and 0.3 ml water to the test sample. Treat the blank solution (BIIIC) in exactly the same way.

ANALYSIS

Using the blank solution (BIIIC), set the Spectronic 20 to read zero absorbance at 450 nm. Measure the absorbance of the test solution IIIC. Record on data page.

Nitrogen Dioxide Determination

Purpose: To determine whether nitrogen dioxide is present in the atmosphere and its concentration.

Equipment:

- 1 10-ml graduated cylinder
- 2 1-liter volumetric flasks
- 3 1-ml transfer pipets
- 10 15-cm test tubes
- 2 100-ml volumetric flasks
- 2 10-ml graduated pipets
- 1 50-ml buret
- Spectronic 20 instrument

Chemicals (Per Student):

Sodium nitrite, $NaNO_2$	0.5 g
10% hydrochloric acid, HCl	22 ml
1% sulfanilimide solution (in HCl)	12 ml
0.1% N(1-naphthyl) ethylenediamine dihydrochloride (aqueous)	12 ml

Outline of Procedure

This procedure is similar to the sulfate analysis in that again there are three parts: 1) the glass fiber filter is exposed to the atmosphere 2-4 weeks; 2) then a series of solutions of known concentration of nitrite ion are treated with a reagent to produce a highly colored solution from which a calibration curve can be formed; and 3) lastly, the residues in the glass fiber are removed as given in the procedure (Part III, test solution (IIIA).

PART IV. PREPARATION OF TEST DETECTION PLATE

Omit this section if the plate has already been prepared for the sulfate determination. If it has not been prepared, follow directions given in Part I of the sulfate determination p. 224.

PART V. PREPARATION OF THE SOLUTIONS FOR THE CALIBRATION CURVE *

a. Preparation of solution VA

Prepare an aqueous standard solution containing 0.493 g dried sodium nitrite, $NaNO_2$ in a 1-liter volumetric flask.

b. Preparation of solution VB

Remove 1 ml of solution VA and dilute to the mark in a 1 liter volumetric flask. What is the concentration of NO_2^- in this solution in grams/liter?

c. **Dilute solution VB further according to the table following using the 1-ml transfer pipets, the 10-ml graduated pipet and the 50-ml buret.**

test tube number	ml $NaNO_2$ (VB)	ml H_2O	ml 10% HCl**	ml 1% sulfa**	ml 0.1% naph-ED**	total vol
(blank)	0	10	1	1	1	13
2	2	8	1	1	1	13
3	4	6	1	1	1	13
4	6	4	1	1	1	13
5	8	2	1	1	1	13
6	10	0	1	1	1	13

** Notes:

1. The 10% hydrochloric acid is prepared by putting 270 ml concentrated acid in a 1-liter volumetric flask and diluting to the mark with deionized water.

2. 1% sulfanilimide solution is prepared by putting 1 g sulfanilimide (sulfa) in a 100-ml volumetric flask and diluting to the mark with 10% HCl.

3. Before adding the 1 ml N(1-naphthyl) ethylenediamine dihydrochloride, let the solution stand 10 minutes after mixing well. This reagent is prepared by putting 0.1 g in 100-ml volumetric flask and diluting to the mark with water. After this solution has been added to the dilution sequence, let the solutions stand about 20 minutes so that the color develops fully.

Set the Spectronic 20 at 580 nm and adjust the instrument to read zero absorbance with the blank (test tube #1). Read and record the absorbances of the remaining solutions.

* See footnote p. 225.

PART VI. PREPARATION OF THE TEST SAMPLE VIA

Dilute 2 ml of solution IIIA to 10 ml in a graduated cylinder with deionize water. To this solution add 1 ml 10% HCl and 1 ml 1% sulfanilimide. Shake well and let stand 10 minutes. Then add 1 ml 0.1% N (1-naphthyl) ethylenediami dihydrochloride solution and let stand 20 minutes. Simultaneously, prepare the blank solution (BVI) in exactly the same way.

ANALYSIS

Set the Spectronic 20 at 580 nm and, using the blank solution (BVI) from Pa adjust the instrument to read zero absorbance. Then measure the absorbance of the test solution VIA.

Chloride Determination

Purpose: To determine the presence and concentration of chloride ion in th atmosphere.

Equipment:

- 1 1-liter volumetric flask
- 6 15-cm test tubes
- 1 5-ml graduated pipet
- 4 100-ml volumetric flasks
- 1 100-ml graduated pipet
- 1 1-ml transfer pipet
- 1 2-ml transfer pipet

Chemicals (Per Student):

0.2 g sodium chloride, $NaCl$
12 ml 1% nitric acid, HNO_3
0.6 g mercury(II) thiocyanate, $Hg(SCN)_2$
13 g ferric alum, $FeNH_4(SO_4)_2 \cdot 12H_2O$

Outline of Procedure

The analysis again consists of 3 parts: Part VII, the collection of the sample; Part VIII, the preparation of a calibration curve, and Part IX, the preparation of the unknown sample.

PART VII. PREPARATION OF THE TEST DETECTION PLATE

See Part I, p. 224.

PART VIII. PREPARATION OF THE SOLUTIONS FOR THE CALIBRATION CURVE *

a. Solution VIIIA

Place 0.1647 g of dried sodium chloride in a 1-liter volumetric flask and dilute to the mark with deionized water. Record weights on data page.

b. Solution VIIIB

Put 10 ml of solution VIIIA into a 100-ml volumetric flask and dilute to the mark with water. What is the concentration of this solution in grams/liter

* See footnote p. 225.

c. **Dilute solution VIIIB further according to the following table using appropriate pipets and burets.**

test tube number	ml NaCl soln. (VIIIB)	ml H_2O	ml 1% HNO_3	ml $Hg(SCN)_2$	ml ferric alum	total vol
1 (blank)	0	20	1	2	4	27
2	2	18	1	2	4	27
3	4	16	1	2	4	27
4	6	14	1	2	4	27
5	8	12	1	2	4	27
6	10	10	1	2	4	27

Notes:

1. Preparation of 1% HNO_3: dilute 1.5 ml concentrated HNO_3 to 100 ml in a volumetric flask.

2. Put 0.6 g $Hg(SCN)_2$ in a volumetric flask. Dilute to 100 ml with methanol.

3. Put 13 g $FeNH_4(SO_4)_2 \cdot 12H_2O$ in 50 ml H_2O. Add 38 ml concentrated HNO_3 and dilute in a volumetric flask to 100 ml with H_2O.

Shake the above tubes (1-6) and let them stand 30 minutes.

Set the Spectronic 20 at 460 nm and adjust the instrument to read zero absorbance with the blank (tube #1). Measure and record the absorbances of the remaining solutions.

PART IX. PREPARATION OF THE TEST SAMPLE

Dilute 7 ml of solution IIIA with 13 ml of deionized water. Prepare a blank solution (BIX) from the solution of the filter from the unexposed plate. Add to both solutions 1 ml 1% HNO_3, 2 ml $Hg(SCN)_2$, and 4 ml ferric alum solution. Let stand 30 minutes. This is solution IX.

ANALYSIS

Use the blank solution (BIX) to adjust the Spectronic 20 to read zero absorbance at 460 nm. Measure the absorbance of the test solution. Record.

NAME ____________________ SECTION ____________ DATE ____________

SPECTROPHOTOMETRIC ANALYSIS OF AIR POLLUTANTS

DATA

II: Sulfur Dioxide Determination

A. What is the SO_4^{2-} concentration in grams/liter of solution IIB? ____________

B. Calculate the number of grams of SO_4^{2-} in each of the dilutions for the calibration curve below and enter the values in the table.

C. Enter the measured absorbance values in the table.

Test tube	ml Na_2SO_4	ml 3% H_2O	ml 1.0 M HCl	ml 0.4 M $BaCl_2$	ml H_2O	Total ml sol.	g SO_4^{2-}	Absorbance 450 nm
1 (blank)	0	1	1	1	18	21		
2	4	1	1	1	14	21		
3	8	1	1	1	10	21		
4	12	1	1	1	6	21		
5	16	1	1	1	2	21		

D. On graph paper plot the grams of sulfate ion on the y-axis against absorbance on the x-axis and draw the best straight line starting at the origin.

E. Write balanced equations for the following chemical reactions:

1. SO_3^{2-} and H_2O_2

2. SO_4^{2-} and Ba^{2+}

III. A. Absorbance of the solution IIIC measured against the blank solution ____________

B. Location of the fiber filter during exposure ____________

Calculations:

1. The absorbance of test solution IIIC ________

2. The number of grams of sulfate ion from the calibration curve ________

3. The number of grams of sulfate in the original sample (before dilution)

$$g\ SO_4^{2-} = g\ SO_4^{2-} \frac{(50)}{7} \frac{(21)}{7}$$

____

4. The number of days the plate was exposed ________

5. Calculate the area of the fiber ________

6. Calculate the number of grams of $SO_4^{2-}/cm^2/day$ ________

7. Convert grams of $SO_4^{2-}/cm^2/day$ to $\mu g\ SO_2/cm^2/day$ ________

$$\frac{(64\ g\ SO_2)(g\ SO_4^{2-})(10^6\ \mu g)}{(96\ g\ SO_4^{2-})(cm^2)(day)(g)} = \_\_\_\_\_\_\_\_ \mu g\ SO_2/cm^2/day$$

8. Possible source of pollutant. ________________________

V. Nitrogen Dioxide Determination

A. What is the NO_2^- concentration in grams/liter of solution VB? ________

B. Calculate the grams of NO_2^- in each of the dilutions of the calibration curves and enter the values in the table below.

C. Enter the measured absorbance values for each solution in the table below.

Test tube number	ml $NaNO_2$ (VB)	ml H_2O	ml 10% HCl	ml 1% sulfa.	ml 0.1% naph. ED	total sol. vol. ml	g NO_2^-	Absorbance 580 nm
1	0	10	1	1	1	13		
2	2	8	1	1	1	13		
3	4	6	1	1	1	13		
4	6	4	1	1	1	13		
5	8	2	1	1	1	13		
6	10	0	1	1	1	13		

D. On graph paper plot the grams of NO_2^- on the y-axis vs absorbance on the x-axis for each of the dilutions. Draw the best straight line through the points starting at the origin.

NAME ______________________ SECTION ______________ DATE ______________

VI. A. Absorbance of test solution VIA ________

B. Location of the fiber filter during exposure ________________

Calculations

1. Absorbance of solution VIA ________

2. Grams of NO_2^- from the calibration curve 

3. Number of grams of NO_2^- in the original samples (before dilution)

$$g\ NO_2^- = \frac{g\ NO_2^-(50)}{(2)}$$ 

4. Number of days the plate was exposed ________

5. Calculate the area of the fiber filter. ________

6. Calculate the number of grams of $NO_2^-/cm^2/day$ ________

7. Convert the grams of $NO_2^-/cm^2/day$ to $\mu g\ NO_2^-/cm^2/day$ ________

8. Possible source of pollutant ________________

VIII. Chloride Determination

A. What is the concentration of chloride ion in grams/liter in solution VIIIB? ____________

B. Calculate the grams of Cl^- in each of the test samples of the calibration curve below.

C. Enter the measured absorbance values for each solution in the table below.

test tube number	ml NaCl (VIIIB)	ml H_2O	ml 1% HNO_3	ml $Hg(SCN)_2$	ml ferric alum	total vol.	g Cl^-	Absorbance 460 nm.
1	0	20	1	2	4	27		
2	2	18	1	2	4	27		
3	4	16	1	2	4	27		
4	6	14	1	2	4	27		
5	8	12	1	2	4	27		
6	10	10	1	2	4	27		

D. On graph paper plot the grams of Cl^- on the y-axis vs absorbance on the x-axis for each solution. Draw the best straight line through the points starting at the origin.

E. Write balanced equations for the following chemical reactions:

1. $Hg(SCN)_2$ and Cl^-

2. Fe^{3+} and SCN^-

IX. A. Absorbance of the test solution IX ________

B. Location of the fiber filter during exposure ________________

Calculations

1. The absorbance of the test solution IX ________

2. The number of grams of Cl^- from the calibration curve ________

3. The number of grams of Cl^- in the original sample (before dilution)

$$g\ Cl^- = \frac{(g\ Cl^-)(50)}{(7)}$$ ________

4. The number of days the plate was exposed ________

5. Calculate the area of the fiber filter. ________

6. Calculate the number of grams of $Cl^-/cm^2/day$ ________

7. Calculate the number of μg of $Cl^-/cm^2/day$ ________

8. Possible source of pollutants ________________

THOUGHT

1. In the determination for SO_2 a student added hydrogen peroxide and barium chloride as indicated and obtained a dense white precipitate.

   a. What had the student failed to do in preparing his sample for analysis?

NAME____SECTION____________DATE____________

b. What might the precipitate be?

2. What is the purpose of

a. the blank solution used in the preparation of the calibration curve (test tube #1 of calibration curve determinations)?

b. the blank solution test sample prepared at the same time as the solution prepared from the test plate? (See IIIA)

Questions 3, 4, 5 may be optional

3. In part II a calibration curve is drawn from the measurements of the absorbances of the known concentrations of the sulfate ion. The concentrations in the directions are given in terms of grams per liter for the sodium sulfate, percent for the hydrogen peroxide, and in moles per liter for the barium chloride. Calculate:

a. The number of moles of sodium sulfate represented by 16 ml of the prepared sodium sulfate solution.

b. the number of moles of hydrogen peroxide present in 1 ml of a 3% solution by weight. (assume the density of the solution to be 1 gram per ml)

c. the number of moles of barium chloride in 1 ml of a 0.4 M solution.

d. 1. Write the equation for the reaction between sodium sulfate and barium chloride.

2. Calculate, (based on the above amount of each reactant present and the equations) the maximum number of moles of barium sulfate which could form in tube #5.

4. Referring to part V, calculate:

a. the number of moles of sodium nitrite in 10 ml of the solution prepared in VB.

b. the number of moles of HCl gas in 1 ml of the 10% hydrochloric acid solution. (assume the density of the solution to be 1 gram per ml)

c. the number of moles of sulfanilimide in 1 ml of a 1% solution.

d. the number of moles of N(1-naphthyl)ethylenediamine dihydrochloride in 1 ml of 0.1% solution. (assume the density of the solution to be 1 gram per ml)

e. Refer to the equations on page 223, and calculate the maximum number of moles of the azo dye which could be produced by these reagents.

5. Referring to part VIII, calculate

a. the number of moles of sodium chloride in 10 ml of the solution prepared in VIIIB

b. the number of moles of nitric acid in 1 ml of a 1% solution (assume the density of the solution to be 1 gram per ml)

NAME____SECTION______________DATE_____________

c. the number of moles of mercury(II) thiocyanate in 2 ml of solution

d. the number of moles of iron(III) alum in 4 ml of the prepared solution

e. Consider the equations for the reactions and calculate the maximum number of moles of iron(III) thiocyanate ion which can form based upon the above values.

6. a. Where did you expose your fiber filter?

b. What kind of pollution was detected?

7. What could be the most likely sources of the pollution you detected?

NAME______________________ SECTION__________ DATE__________

Experiment 18
ANALYSIS OF THE IRON CONTENT OF VITAMIN PILLS

PRELIMINARY QUESTIONS

1. How many ppm of iron(II) ions are in 100 ml of an aqueous solution containing 0.01 g of iron(II)? (Assume the density of the solution to be 1 gram per ml.)

2. Write balanced equations for the following reactions:

   a. iron(II) ion and hydroxylamine

   b. 1,10-phenanthroline and iron(II) ion.

Experiment 18
ANALYSIS OF THE IRON CONTENT OF VITAMIN PILLS

IDEAS

In all living species, the molecules which make up the organisms contain a large percentage of carbon, oxygen, hydrogen, and nitrogen. In addition to these four elements, which together account for 99 percent by weight of these molecules, there are twenty more elements essential to life. However, more than half of these are required in only trace amounts.

Diet is the major factor in maintaining good health. In 1912 positive evidence was obtained that certain organic compounds were necessary in the diet of higher animals to cure or prevent such deficiency diseases as scurvy and rickets.* These were named vitamins. Four years later it was recognized that dietary requirements also include various inorganic ions. As early as 1860, L. Pasteur had found that yeast requires inorganic ions in the culture medium.

Among the most important of inorganic cations is iron, which is indispensible to higher animals in small or trace amounts. The iron content of the body is about 4.3 grams per 70 kg or about 0.006 percent. Iron is found in the body as a component of hemoglobin. Fifty-five percent of the total iron in the body is tied up in this complex species. Of the remainder, 35 percent is stored iron called ferritin.

The human body must have iron in sufficient amounts in order to form the hemoglobin which acts as a transporter of oxygen. The hemoglobin in its red form releases the oxygen to the tissues for cell metabolism as the blood flows through the body. If there is a deficiency of iron, the body's cells produce insufficient energy leading to symptoms of sluggishness, drowsiness and anemia. Many formulations of vitamin supplement tablets therefore contain small amounts of iron.

In this experiment the quantitative determination of iron is made colorimetrically using the spectrophotometer. The analysis depends on the formation of a deeply colored complex. Read the introductory section on spectrophotometric methods before performing this experiment.

There are two steps in the analysis:

1) hydroxylamine hydrochloride ($H_2NOH{\cdot}HCl$) is added to the solution to be analyzed to reduce all Fe^{3+} to Fe^{2+}.

$$4\,Fe^{3+} + 2H_2NOH \longrightarrow 4Fe^{2+} + N_2O + 4H^+ + H_2O$$

* Hopkins, F. G., *J. Physiol.*, *44*, 425 (1912)

2) The Fe^{2+} is converted into a highly colored complex with 1, 10-phenanthroline

$$Fe^{2+} + 3(C_{12}H_8N_2) \longrightarrow Fe(C_{12}H_8N_2)_3^{2+}$$

lt. green   no color   orange-red

The semi-structural formula of the complex ion is

Coordination of phenanthroline with Fe^{2+}

A spatial configuration is

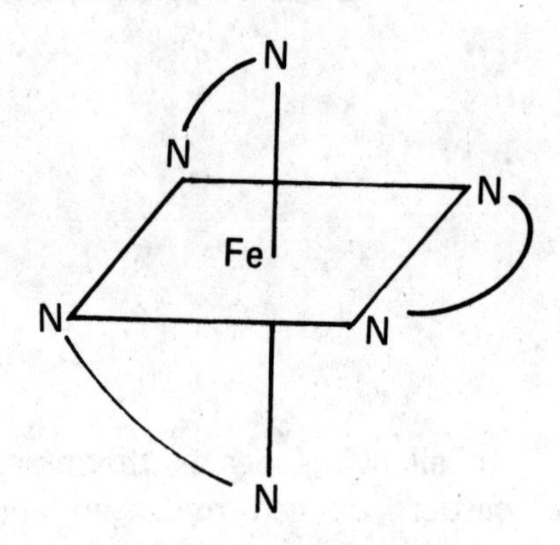

Octahedral structure configuration

The colored solution of the complex will absorb a maximum of light at 508 nm and should have a pH of about 3.5 to prevent the precipitation of various iron salts from the solution.

In analyses for constituents present in small amounts, the concentration is quite frequently given as parts per million (ppm). In this experiment units are based on a weight/weight ratio

$$\text{number of ppm} = \frac{\text{wt constituent}}{\text{wt sample}} \times 10^6$$

If one liter of a solution had 4 ppm in copper ion, the solution would contain 0.004 g copper ion. Since these aqueous solutions are very dilute, the density of the solutions is assumed to be the same as water.

The absorbance of the solutions are measured on the Spectronic 20. Directions for the use of this instrument are given in the experiment titled: Spectrophotometric Determination of the Formula of a Complex Ion.

INVESTIGATION

Purpose: To determine a trace concentration of a metal ion in a commercia vitamin pill by spectrophotometric analysis.

Equipment:

Spectronic 20
3 100-ml volumetric flasks
1 10-ml transfer pipet
1 5-ml transfer pipet
#40 filter paper
1 2-ml transfer pipet
1 1-ml transfer pipet

Chemicals: Standard iron (II) solution (10 ml/student)

Preparation of 1 liter: dissolve 0.2808 g $Fe(NH_4)_2(SO_4)_2 \cdot H_2O$ in 1 liter of solution which contains 2.5 ml concentrated sulfuric acid.

1,10-phenanthroline (0.25%): Dissolve 0.25 g 1,10 phenanthroline in 100 ml of solution containing 10 ml ethanol.

Hydroxylamine hydrochloride, $H_2NOH \cdot HCl$: 10 g in 90 g water.

Sodium acetate, $NaC_2H_3O_2$: saturated aqueous solution.

Assorted commercial vitamin pills containing iron.

Procedure:

I. Preparation of the Standard Complex Solutions for the Calibration Curve

A. 4 ppm Fe^{2+} solution

Using a 10-ml pipet, transfer 10 ml of the standard iron (II) solution into a 100-ml volumetric flask. Add 10 ml of saturated sodium acetate solution which will buffer the solution between pH 3 and pH 6. Add 2 ml of the hydroxylamine hydrochloride solution and 3 ml of the 1,10 phenanthroline solution into the flask. Dilute to the mark with deionized water and mix well.

B. 2 ppm Fe^{2+} solution

First, pipet 5 ml of the standard iron solution into a 100-ml volumetric flask and then follow with directions exactly as described in 'A' above.

C. 0.8 ppm Fe^{2+} solution

Pipet 2 ml of the standard iron solution into a 100-ml volumetric flask and proceed exactly as described in 'A' above.

D. 0.4 ppm Fe^{2+} solution

Pipet 1 ml of the standard iron solution into a 100-ml volumetric flask and proceed exactly as described in 'A' above.

Allow at least 15 minutes after adding the reagents before making absorbance measurements. This gives time for the color of the complex to develop fully.

While you are waiting, prepare a blank solution by putting 2 ml hydroxylammine hydrochloride solution, 3 ml of the 1,10 phenanthroline, 10 ml of the sodium acetate solution in a 100-ml volumetric flask. Add deionized water to the mark and shake well.

Measure the absorbance of your prepared complexed iron (II) solution against the blank at 508 nm.* Record the absorbance values.

---

* The formation of the gas N_2O may cause problems in reading the absorbance accurately. Tap the tube gently to dislodge bubbles.

Record all data on the data page.

II. Preparation of the Calibration Curve Using Beer's Law (A = abC)

Plot the absorbance reading on the y axis versus the concentration of ppm Fe^{2+} in the complexed solutions on the x axis. This plot should be linear and g through the origin. Draw the best straight line possible consistent with the data. The slope of the line is ab. However, if you use 1 cm absorption cells the slope has the value 'a' or the absorptivity.

Record your value for absorptivity on the data page.

III. Determination of the Iron Content of the Vitamin Pill

Place one vitamin pill in a 100-ml beaker. Add 25 ml of 6 M HCl and boil gently for about 15 minutes. Add 10 ml of deionized water and filter the solution immediately through #40 filter paper directly into a 100-ml volumetric flask. Wash the filter paper containing any residue with two or three 5 ml portions of deionized water adding these washings to the volumetric flask. When the contents of the flask have reached room temperature, dilute to the mark with deionized water. Label this solution IIIA.

Pipet a 5 ml aliquot (10 ml if the label indicates that the tablet contains less than 15 mg of iron) into a 100-ml volumetric flask and dilute to the mark. Mix well. Label this solution IIIB.

The unknown iron solution is prepared for analysis by pipeting 10 ml of the solution IIIB into a 100-ml volumetric flask, adding 10 ml of the saturated sodium acetate solution, 2 ml 10% hydroxylammine hydrochloride solution, and 3 m of the 1,10 phenanthroline reagent.

Wait 15 minutes and determine the abosrbance.

IV. Calculation of the Concentration of Iron (II) in the Vitamin

Mark off your absorbance value for the unknown iron (II) solution on the graph and read the corresponding concentration in ppm. Record your reading.

This value represents the concentration of the iron in the diluted solution of the unknown. It is necessary to convert this value to milligrams of iron per tablet. If your tablet supposedly contains 25 mg iron, dissolved in the 100-ml volumetric flask, it contains 250 ppm iron. By diluting 5 ml of this solution 100 ml the iron concentration is reduced by a factor of 20, so that the iron concentration becomes 12.5 ppm. The next dilution is 10 ml to 100 ml (or a 1/10 reduction) so that the iron concentration becomes 1.25 ppm. The equations given below have the function of converting the diluted value to the original weight of the iron in the tablet dissolved in the first 100-ml volumetric flask.

If you used a 5 ml aliquot of the dissolved solution:

$$\frac{\text{mg iron}}{\text{tablet}} = \frac{\text{ppm (as read from graph)}(100\text{ ml})}{(1000\text{ ml})}\frac{(100\text{ ml})(100\text{ ml})}{(5\text{ ml})(10\text{ ml})}$$

If you used a 10 ml aliquot of the dissolved tablet solution:

$$\frac{\text{mg iron}}{\text{tablet}} = \frac{\text{ppm (as read from graph)}}{(1000\ \text{ml})} \frac{(100\ \text{ml})(100\ \text{ml})(100\ \text{ml})}{(10\ \text{ml})\ (10\text{ml})}$$

Record your value calculated from the above equation.

Compare your experimental value with that on the label.

NAME____SECTION_____________DATE____________

ANALYSIS OF THE IRON CONTENT OF VITAMIN PILLS

DATA

I. CALIBRATION DATA

Absorbance readings of standard 1,10 phenanthroline Fe (II) complex

Concentration Fe^{2+} (ppm)	Absorbance readings

II. Absorptivity 'a'
(determined by Beer's law plot) ____________

III. Absorbance of vitamin pill: brand name ______________________

a. Initial aliquot of dissolved vitamin solution taken (5 ml or 10 ml) ____________

b. Absorbance ____________

IV. a. Concentration of Fe^{2+} in ppm as read from graph ____________

b. mg Fe/tablet (show calculations) ____________

c. labelled value of Fe/tablet ____________

d. percent difference (show calculations) ____________

THOUGHT

1. Recommended daily allowance of iron for adults is 10 mg for males and 12 mg for females. The mineral contents of foods can be obtained from Food, the Yearbook of Agriculture, U. S. Department of Agriculture, Washington, D.C. Make a record of the kinds and quantities of food you eat during one day and compute the amount of iron present in these foods.

   Does your diet (of that day) meet or exceed the minimum daily requirement?

2. What is the function of the blank when using the Spectronic 20?

NAME______________________________________SECTION____________DATE__________

Experiment 19

DETERMINATION OF THE FORMULA OF A COMPLEX ION

PRELIMINARY QUESTIONS

1. What is Job's method?

2. a) What is Beer's law?

   b) What are some of the limitations of Beer's law?

3. How do you prepare 100 ml of 0.1 M nickel nitrate hexahydrate?

4. How do you prepare 100 ml of 0.2 M dihydrate of the disodium salt of ethylenediaminetetraacetic acid,* molecular weight of 372.3?**

---

* Because of its limited solubility, ethylenediaminetetraacetic acid is not use in these procedures. The disodium salt of the dihydrate is preferred not onl because of its greater solubility but because it can be easily prepared in a pure form. In this set of experiments the disodium dihydrate salt will be called EDTA.

** The semistructural formula for the molecule is

$$\begin{array}{l} Na^{+}\,{}^{-}OOCCH_2 \diagdown \qquad\qquad\qquad\qquad \diagup CH_2COO^{-}\,Na^{+} \\ \qquad\qquad\qquad N\text{-}CH_2\text{-}CH_2\text{-}N \qquad\qquad\qquad (2H_2O) \\ HOOCCH_2 \diagup \qquad\qquad\qquad\qquad \diagdown CH_2COOH \end{array}$$

Experiment 19

DETERMINATION OF THE FORMULA OF A COMPLEX ION

IDEAS

Discussion of Procedure

The complex for which the empirical formula will be found is that formed by the reaction between nickel ion and the disodium salt of ethylenediaminetetraacetic acid (EDTA).

Both the nickel ion (from the nickel nitrate) and the nickel complex are colored but they absorb light in different regions of the visible spectrum; aqueous solutions of nickel nitrate are green and the solutions of Ni–EDTA complex are blue in color. Therefore, by measuring the absorbances at a specific wavelength as the color of the solutions changes from green to blue, it is possible to determine the formula of the complex. The instrument used to measure the change in color is the Spectronic 20.

The method to be used is called Job's method, the method of continuous variation. It is a procedure in which the concentration of metal ion and ligand is varied systematically. However, the total volume of ligand* and metal ion solution and the sum of their concentrations remains constant. The absorbance of light at a specific wavelength, where the complex absorbs most strongly, is directly proportional to the concentration of the colored species and is measured with the Spectronic 20 for all the solutions. Because the volume of all solutions is the same, a ratio of the concentration of Ni^{2+} to EDTA in moles/liter is, in fact, also the ratio of the number of moles of each.

When the absorbance of the complex is plotted against the ratio of concentration of Ni^{2+} to EDTA, a curve with a maximum is formed. If the curve has a sharp peak, then the reaction is almost complete. If the curve is smooth, that is has a broad maximum, then an equilibrium exists between the complex and a large number of ions. It is possible to extrapolate both legs of this curve to obtain a sharp peaked maximum at this intersection. This sharp peak, either from experimental data or from extrapolation of the curve, gives the molar ratio of nickel to EDTA in the complex.

There are four parts to the experimental procedure:

1. the preparation of solutions to be used for Job's method
2. use of the blank and the selection of the appropriate wavelength
3. the test of Beer's law
4. the application of Job's method.

*Molecules or negative ions surrounding the central ion of a complex ion are called ligands; in this case the EDTA is the ligand and the nickel ion is the central ion.

INVESTIGATION

Purpose: To determine the formula of a complex ion by using Job's method.

Equipment:

2 100-ml volumetric flasks
2 30-ml beakers
11 15-cm test tubes
12 10-cm test tubes or cuvettes
2 10-ml graduated pipets
1 red-sensitive phototube and
1 blue-sensitive phototube
for the Spectronic 20

Chemicals: nickel(II) nitrate hexahydrate, $Ni(NO_3)_2 \cdot 6H_2O$
EDTA, dihydrate disodium salt, solid

Procedure:

I. Solutions for Job's Method

Weigh the amount of nickel nitrate needed to prepare 100 ml of 0.1 M solution in a weighed clean dry beaker. The weights need to be known only to the nearest .01g. Carefully transfer the nickel salt to a 100-ml volumetric flask using several rinsings of small amounts of deionized water to ensure that no nickel nitrate has remained in the beaker(*)*. Record on the data page.

Calculate the amount of EDTA needed to prepare 100 ml of 0.1 M solution, and prepare this solution by the method described above (*). Record on the data page.

Put eleven 15-cm test tubes in a rack and number them serially. Then proceed according to the chart on the data page to mix the indicated volumes together in each test tube. Use a graduated pipet to measure the volumes.

After the solutions are mixed, several observations and inferences can be made:

1. Nickel nitrate solution is greenish in color and probably contributes some color to those solutions in which it is present in excess.
2. The most intense blue color is found in test tube 6.
3. EDTA alone in aqueous solution is colorless, so that decrease in color intensity in test tubes 7-10 is probably due to a dilution of the blue species.

II. Use of the Blank and Wavelength Selection

The next step is to find the wavelength of light which will give the maximum absorbance for this blue solution (test tube 6).

Examine the Spectronic 20. It should contain a blue-sensitive phototube. Check with your instructor. Notice that the dial on top front reads both absorbance and transmittance directly. One hundred percent transmittance is zero absorbance and infinite absorbance is zero transmittance. The absorbance scale

---

* Convenient stopping places are marked by asterisks within parenthesis (*) indicating the experiment may be interrupted at this point for an extended period.

is logarithmic, hence few significant figures can be obtained at high absorbance. The Spectronic 20 functions most precisely in the middle ranges of the absorbance scale from about 0.1 to 0.7 (80-20 percent transmittance). Turn on the power switch and zero control knob on the left front of the spectrophotometer and allow the instrument to warm up at least 20 minutes. Then set the wavelength by turning the dial on top right to 360 nm and adjust the zero control so that the pointer on the transmittance-absorbance scale reads zero transmittance.

Obtain two 10-cm test tubes. Using very small labels, label one H_2O, the other 6*. Attach the labels just below the lip. Draw a vertical line on each label. Fill each test tube about one-half full with the indicated solution. The length of the optical path and the medium through which the light passes are critical in making these measurements. Therefore, it is important that the test tubes always be inserted in the sample holder in exactly the same way. If you open the top of the sample holder, you will see a line (ridge) on the front of the opening. When you insert your test tube, be sure to line up this ridge on the holder with the vertical line on your label. It is also important that the test tube be clean. Wipe the outside with lens paper to remove fingerprints. Obviously any smudge will alter the light path and hence the results.

Insert your 10-cm test tube containing the water into the sample holder of the Spectronic 20 and adjust the light control (knob on the right front) so that the pointer on the scale reads zero absorbance. Remove the sample and transfer to a 10-cm test tube some of the solution labeled 6. Read the absorbance of this solution on the scale. Record the following data in order to select the wavelength of maximum absorbance of the complex.

Starting with 360 nm measure the absorbance of your solution at 20 nm wavelength increments.** Each time the wavelength is changed, the following procedure must be adopted:

1. Adjusting the zero point

   The zero point with no sample in the closed sample holder must be adjusted to zero transmittance, an adjustment made by using the zero control knob. Be sure to *close* the sample holder for all readings. (When no sample is present, a mechanical barrier prevents light from striking the photocell.)

2. Use of the blank

The test tube containing the water must be inserted in the sample holder and the absorbance adjusted to zero (100 percent transmittance) using the light control knob. Since all substances absorb some light, the effect of the absorbance of light by the solvent is eliminated by adjusting the instrument to

---

* Optical cuvettes may be used if they are available. However, regular test tubes do not offer any problems if they are clean, free from surface scratches, round, and produced by the same manufacturer. Check manufacturer's marks.

** If the needle is deflected to the far right off the scale, *turn the light source down immediately*.

read zero absorbance with a sample of the pure solvent. Its effects are thereby screened out. This procedure establishes a blank.

The test tubes must be inserted in the holder in exactly the same position each time. Wipe the outside surfaces of the lower half of the test tubes with lens paper and check them carefully to be sure there are no smudges or fingerprints on them.

Continue reading the absorbance as the wavelength is increased in 20 nm increments to 600 nm.

Examine your data to find an approximate maximum value for absorbance. Wher this value has been found, the next step is to locate the maximum absorbance more precisely. This is done by measuring the absorbance in 5 nm increments, 10 nm before and after what you have determined as the approximate maximum value. (*

On the data page, plot on graph paper the absorbance on the y-axis and the wavelength in nanometers on the x-axis. Connect the points with a smooth line. Determine the wavelength at which the complex absorbs the most light.

III. The Test of Beer's Law

Beer's law states that absorbance is directly proportional to the concentra tion of the absorbing species in the solution. Therefore, a plot of absorbance against concentration where concentrations are varied should give a straight line. There are several experimental factors which must be considered in order to obtain good results.

1. Incident light is both reflected and absorbed by the glass. To compensate for this, be sure to use the same test tube to hold the samples of the complex and always insert it in the sample holder in exactly the same way. Inciden light can also be absorbed unequally or scattered by heterogeneous dispersions. Gas bubbles escaping from the solution may also cause trouble. These effects ca be minimized by proper mixing of the solutions and avoiding concentrated solutions.

2. Beer's law will not hold if the concentration of absorbing species cha as a result of the dilution. Often a change in pH will also cause this to occu The complex which we are investigating seems to be stable at the concentrations

Using graduated pipets, make up 20 ml of Ni-EDTA complex by mixing 10 ml o 0.1 M EDTA with 10 ml 0.1 M $Ni(NO_3)_2$ in a 15-cm test tube. Insert a rubber stopper and invert at least ten times to mix thoroughly. Again using graduated pipets, dilute the sample as follows and put each into a clearly labeled 10-cm test tube.

	Test Tube
1 ml Ni-EDTA complex + 4 ml H_2O	1
2 ml Ni-EDTA complex + 3 ml H_2O	2
3 ml Ni-EDTA complex + 2 ml H_2O	3
4 ml Ni-EDTA complex + 1 ml H_2O	4

The original sample will be called test tube 5.

Record the concentration of the original sample and the dilutions on the data page.

Using the blue-sensitive phototube and a wavelength of maximum absorbance, set the transmittance to read zero when no sample is in the sample holder. Insert the deionized water sample into the sample holder and set the *absorbance* at zero.

Measure the absorbance of each of the above samples. For each sample, be sure the zero point is set at zero transmittance when no sample is used and at zero absorbance when the sample of deionized water is tested. For each measurement of absorbance on samples 1-5, use the *same* 10-cm test tube or cuvette, and insert in the sample holder in the same position each time. After each test return the sample to the test tube from which it was removed, rinse the test tube used in the measurement twice with 1/2-ml increments of the next solution to be measured, and discard the rinsings. Wipe the outside of the test tube with lens paper to remove fingerprints and smudges and insert the test tube half full of the test sample. Record the absorbance of each solution.

Using graph paper plot absorbance on the y-axis and concentration on the x-axis. With a ruler, and starting at the origin, draw the best straight line consistent with experimental points.

IV. Job's Method: Continuous Variation of Concentration

Set the wavelength as in section III. Establish zero transmittance when no sample is in the holder, and zero absorbance with deionized water in the previously described manner. Because the nickel nitrate solution is green, it alters the absorbance of the solutions in which it is present in excess. The amount of this absorbance will affect the shape of the curve but will have no effect on the location of the maximum in the curve. Therefore, it is not necessary to determine the effect of this species on the absorbance of the solutions. Measure the absorbance of each of the solutions in the 15-cm test tubes prepared in (I). Use the 10-cm test tubes to hold the samples as previously described. Record results in Table IV. Plot graphs on data page.

NAME____________________ SECTION__________ DATE__________

DETERMINATION OF THE FORMULA OF A COMPLEX ION

DATA

Weight of beaker and $Ni(NO_3)_2 \cdot 6H_2O$ __________

Weight of beaker __________

Weight of $Ni(NO_3)_2 \cdot 6H_2O$ __________

Weight of beaker and EDTA __________

Weight of beaker __________

Weight of EDTA __________

I. Solutions for Job's Method

Test Tube	0.10 M $Ni(NO_3)_2$ (ml)	0.10 M EDTA (ml)
1	10	0
2	9	1
3	8	2
4	7	3
5	6	4
6	5	5
7	4	6
8	3	7
9	2	8
10	1	9
11	0	10

II. Selection of Proper Wavelength Using Test Tube 6

Wavelength, (nm)	Absorbance, A	Blue-sensitive phototube
360		
380		
400		
420		
440		
460		
480		
500		
520		
540		
560		
580		
600		

Wavelength range indicated by approximate values for absorbance ________

Wavelength, 5 nm increments	Absorbance, A

Wavelength of maximum absorbance______

NAME ______________________ SECTION ____________ DATE ____________

On graph paper, plot absorbance on the y-axis and wavelength on the x-axis.

Which wavelength gives the maximum absorbance? ____________

III. Test of the Beer's Law

Assume the original concentration of the Ni-EDTA complex is 0.05 M

Solution	test tube	Conc. Ni-EDTA	A
1 ml Ni-EDTA complex + 4 ml H_2O	1		
2 ml Ni-EDTA complex + 3 ml H_2O	2		
3 ml Ni-EDTA complex + 2 ml H_2O	3		
4 ml Ni-EDTA complex + 1 ml H_2O	4		
original sample	5		

On graph paper, plot the absorbance on the y-axis and the concentration of the Ni-EDTA on the x-axis. With a ruler, and starting at the origin, draw the best straight line consistent with experimental points.

How well do your results reflect Beer's law?

IV. Job's Method

Test tube	0.1 M $Ni(NO_3)_2$ (ml)	0.1 M EDTA (ml)	Moles Ni^{2+}	Moles EDTA	Absorbance A
1	10	0			
2	9	1			
3	8	2			
4	7	3			
5	6	4			
6	5	5			
7	4	6			
8	3	7			
9	2	8			
10	1	9			
11	0	10			

NAME____ SECTION____________ DATE__________

Plot the number of moles of Ni^{2+} and EDTA on the x-axis and absorbance, A, on the y-axis.

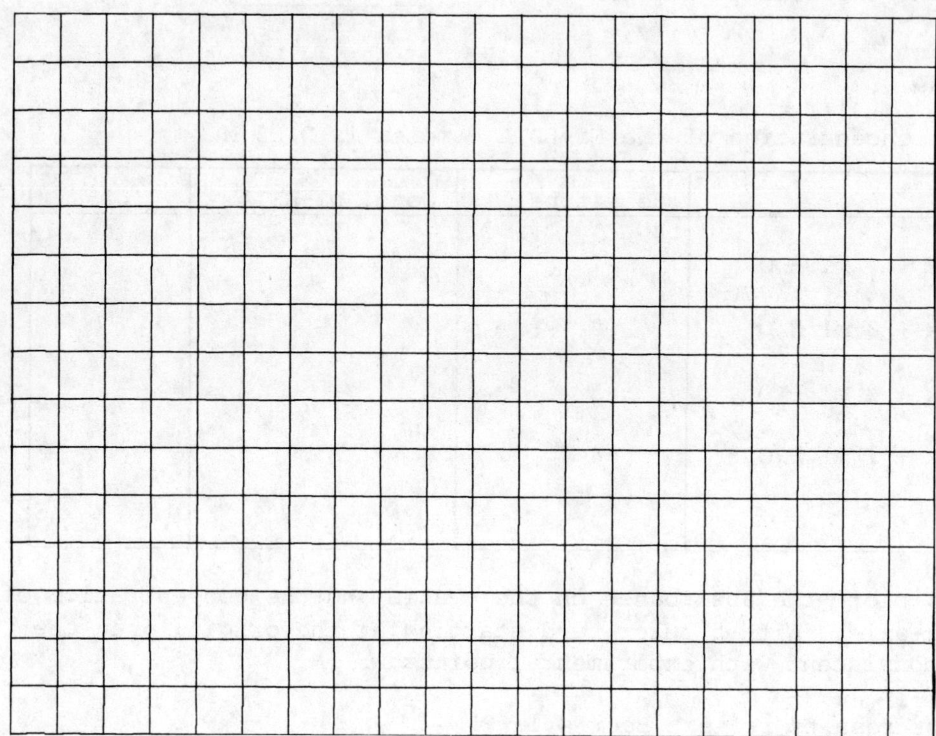

moles Ni^{2+} (decreasi
moles EDTA (increasi

What is the empirical formula for this complex?

THOUGHT

1. Why is it necessary to wipe carefully the outside surface of the cuvette or test tube before insertion into the spectrophotometer?

2. Why is the pure solvent used to set the Spectronic 20 to read zero absorbanc before other readings are taken?

3. Why is it necessary to insert the test tubes or cuvettes always with the same alignment?

4. What would the Job's method curve look like if the ratio of ligand to metal ion were 3:1?

NAME____SECTION______________DATE__________

Experiment 20

DETERMINATION OF AN EQUILIBRIUM CONSTANT

PRELIMINARY QUESTIONS

1. Define the following terms:

   a) catalyst

   b) reactant

   c) product

   d) concentration

   e) rate of a chemical reaction

   f) esterification

   g) hydrolysis of an ester

2. a) Write the formula for acetic acid.

   b) Write the formula for ethyl alcohol.

3. How would you prepare 100 ml of 1 N hydrochloric acid from 12 N hydrochloric acid?

4. a) Write the equation for the hydrolysis of ethyl acetate.

   b) Write the equilibrium constant expression for the hydrolysis of ethyl acetate.

5. a) Write the equation for the reaction between ethyl alcohol and acetic acid.

   b) Write the equilibrium constant expression for the synthesis of ethyl acetate.

Experiment 20

DETERMINATION OF AN EQUILIBRIUM CONSTANT

IDEAS

The search to discover the reasons for chemical union stimulated a great deal of thought and experimentation for hundreds of years. Why do iron and oxygen combine to form rust, while gold resists such union? As early as the thirteenth century, Bishop Magnus, using ideas borrowed from Hippocrates, explained chemical action as occurring because of the "affinity" of the reacting substances for each other due to their similarity. In the early nineteenth century, H. Davy, J. Berzelius, and M. Faraday (see Ideas, Experiment 12) suggested that this affinity was electrical in nature. C. Berthollet (French physician, chemist, friend of Napoleon, and J. Proust's polite adversary in their controversy on definite proportions) studied chemical affinity and proposed that an important role is played by the masses of reacting substances. A reaction may proceed further if the concentration, mass per unit volume, of the reactants is increased, or the course of the reaction may be reversed by adding sufficient quantities of the products. The concepts of equilibrium and reversible reactions were born here.

Investigators carefully studied the rate of a chemical reaction in 1850. In 1855 J. Gladstone investigated the course of a reaction probably familiar to you: the addition of potassium thiocyanate to ferric salts, which produces varying shades of red, depending upon the concentration of the products. In 1862 M. Berthelot and P. St. Giles, in a series of careful studies of the reactions of acids and alcohols (acid + alcohol → ester + water), confirmed earlier ideas that the mass of the reactants directly influences both the speed of the reaction and the concentrations of the products. The reaction between the alcohol and the acid is never complete, but appears to stop after a certain concentration of ester and water determined by experimental conditions has been produced.

Ideas on the influence of the concentration of the reactants on the speed and products of a chemical reaction were combined in 1864 by the Norwegians C. Guldberg (1836-1902) and P. Waage (1833-1900) in their great mathematical generalization, the law of mass action. The velocity of a reaction at constant temperature is proportional to the product of the active masses of the reacting substances. They wrote, "We must study the chemical reactions in which the forces which produce new compounds are held in equilibrium by other forces . . . where the reaction is not complete but partial."* Here is the idea of a dynamic chemical equilibrium rather than a static equilibrium, a dynamic equilibrium which is produced by the attaining of equal speeds by the forward and backward reactions. The mathematical expression for the law of mass action that is accepted now evolved from the ideas of Guldberg and Waage.**

* Etudes sur les affinites chimiques, Christiana, 1867; Ostwald's *Klassiker* No. 104; quoted in J. R. Partington, *A Short History of Chemistry*, 3rd ed., Harper & Bros., New York, 1957, p. 326.

** See E. Lund, "Guldberg and Waage and the Law of Mass Action," *J. Chem. Educ.*, 42, 548-550 (1965).

It may be illustrated by the following:

$$aA + bB + \ldots \rightleftarrows cC + dD + \ldots$$

$$K_{eq} = \frac{[C]^c[D]^d}{[A]^a[B]^b}$$

where A, B, C, D represent reactants and products and a, b, c, d the coefficients necessary to balance the equation. Square brackets are used to indicate concentration, in moles per liter.

The attainment of chemical equilibrium, as has just been noted, is determined by the rates of the forward and backward reactions. The study of reaction rates is therefore obviously closely related to the study of chemical equilibria.

Reaction rate is the speed with which a reaction occurs. If we assume the coefficients necessary for balancing the general equation above are one, we can rewrite it as follows:

$$A + B \rightleftarrows C + D$$

If the reaction is read from left to right, A and B are reactants and C and D are products. The double arrow, signifying a reversible reaction, indicates, however, that C and D can also be reactants, and A and B products:

$$C + D \rightleftarrows A + B$$

Both forward and reverse reactions occur simultaneously. Reaction rates are affected by the concentration of the reactants and products, temperature of the reaction, nature of the products and reactants (ease with which bonds are broken and formed), and whether a catalyst is present. To illustrate, assume that reactants A and B are placed in a flask. The initial reaction is between a molecule of A and a molecule of B to form a molecule of C and a molecule of D. Once these products have formed they may react to the form A and B. The rate initially between reactants A and B will be most rapid. As the products form and the concentrations of A and B decrease, the forward reaction rate slows down (unless the reaction is autocatalytic). However, the reverse reaction rate between C and D to form A and B increases as the concentration of C and D increases.*

To calculate a reaction rate, careful measurements are made to determine the decrease in concentration of a reactant at specified time intervals. Since many reactions between inorganic compounds are extremely fast because of the ionic nature of the substances, a study of their reaction rates is difficult. Consequently, a chemical reaction in which organic substances are used, where covalent bonds are broken and formed, will be studied in this experiment.

$$\underset{\text{ethyl alcohol}}{CH_3CH_2OH} + \underset{\text{acetic acid}}{CH_3\overset{\overset{O}{\|}}{C}{-}OH} \overset{H^+}{\rightleftarrows} \underset{\text{ethyl acetate}}{CH_3\overset{\overset{O}{\|}}{C}{-}O{-}CH_2CH_3} + \underset{\text{water}}{H_2O}$$

At room temperature this reaction takes several hours, even when the rate is accelerated with a catalyst, H^+. The rate of the forward reaction can easily be determined by measuring the decreasing concentration $\left(\frac{\text{moles}}{\text{liter}}\right)$ of acetic acid.

---

* Although the stoichiometric equation gives an outline of the overall reaction, the mechanism of the reaction is complicated and often unknown. Rates are always measured experimentally. For further information, consult a text on the mechanics of chemical reactions.

The rate of the reverse reaction, hydrolysis of the ester, is measured by determining the increase in concentration of acetic acid with time. In this case the increase in acid is a measure of the decrease in concentration of the ester.

Equilibrium in this or any reaction is achieved when the rate of the forward reaction is equal to the rate of the reverse reaction (when the rate with which A and B react to form C and D is equal to the rate at which C and D react to form A and B).

By definition an equilibrium constant is the product of the concentrations of the products divided by the product of the concentration of the reactants.

$$\frac{[C]\,[D]}{[A]\,[B]} = K_{eq}$$

The only factor which can change the value of the equilibrium constant is a change in temperature.

The equilibrium constant expression for the hydrolysis of ethyl acetate is

$$\frac{[CH_3\overset{O}{\overset{\|}{C}}OH]\,[CH_3CH_2OH]}{[CH_3\overset{O}{\overset{\|}{C}}{-}OCH_2CH_3]\,[H_2O]} = K_{eq}$$

The expression for the synthesis of ethyl acetate is

$$\frac{[CH_3\overset{O}{\overset{\|}{C}}{-}OC_2H_5]\,[H_2O]}{[CH_3\overset{O}{\overset{\|}{C}}OH]\,[CH_3CH_2OH]} = K_{eq}$$

Note that these expressions are derived from the equation on page 266. Mathematically one expression is the reciprocal of the other. The hydrogen ion added as a catalyst remains constant in concentration throughout the reaction and need not be included in the expression.

INVESTIGATION

Purpose: To plot a reaction rate curve; to calculate the value of the equilibrium constant from experimental data for reactions between acetic acid and ethanol, and the hydrolysis of ethyl acetate.

Equipment:

- 6 15-cm test tubes with tight-fitting corks, and labels
- 4 transfer pipets: 10 ml, 5 ml, 2 ml, 1 ml
- 1 50-ml buret
- 1 125-ml Erlenmeyer flask
- 1 500-ml Florence flask with rubber stopper

Chemicals:

- ethyl acetate, $CH_3\overset{O}{\overset{\|}{C}}{-}OC_2H_5$
- ethyl alcohol, C_2H_5OH
- glacial acetic acid, $CH_3\overset{O}{\overset{\|}{C}}{-}OH$
- 12 N hydrochloric acid, concentrated
- sodium hydroxide, NaOH, solid
- 0.1% solution of phenolphthalein

Procedure:

I. a. Prepare 100 ml of approximately 1 N hydrochloric acid from 12 N hydrochloric acid by dilution.

b. Label the six 15-cm test tubes c1, c2, c3, and d1, d2, d3.

c. In each of the three test tubes labeled c1, c2, c3 put 10 ml of ethyl acetate, 2 ml of water, and 1 ml of 1 N hydrochloric acid; measure these precisely with a pipet.* Record the time.

d. In each of the three test tubes labeled d1, d2, d3 pipet 5 ml of glacial acetic acid, 5 ml of ethyl alcohol, and 1 ml of 1 N hydrochloric acid. Record the time.

e. Stopper the test tubes and allow to stand in a beaker or test tube rack.

f.

Test tube	Contents	Date and time
c1 c2 c3	10 ml $CH_3COOC_2H_5$ 2 ml H_2O 1 ml 1 N HCl	
d1 d2 d3	5 ml glacial CH_3COOH 5 ml C_2H_5OH 1 ml 1 N HCl	

II. Prepare 500 ml of 2 N sodium hydroxide. Standardize this solution against glacial acetic acid in the following way:

Carefully pipet* 5 ml of glacial acetic acid (density 1.05 $\frac{g}{ml}$) and transfer to a 250-ml Erlenmeyer flask. Dilute the acid with about 50 ml of deionized water, add 3 drops of phenolphthalein indicator, and titrate with the 2 N sodium hydroxide to a pale pink endpoint. Repeat the titration until three reproducible values for the value of the normality of the s[odium] hydroxide are obtained (two parts per hundred). Read the volume on the bure[t] to the nearest 0.02 ml.

Calculate the normality to three significant figures of the sodium hydroxide as follows:

number of equivalents of acetic acid = number of equivalents of sodium hydroxide

$$N = \frac{\text{number of equivalents}}{\text{liter}}$$

$$(5 \text{ ml acid})\frac{(1.05 \text{ g})}{(\text{ml})} \frac{(\text{number of eq})}{(60 \text{ g})} = (\text{ml base})(\text{N base})\frac{(\text{liters})}{(10^3 \text{ ml})}$$

* See Laboratory Techniques for methods of pipetting.

III. Pipet 20 ml of the approximately 1 N hydrochloric acid (preparation described in Part Ia) into a 125-ml Erlenmeyer flask, add 3 drops of phenolphthalein, and titrate with your 2 N sodium hydroxide. Repeat this titration again until three reproducible values are obtained (two parts per hundred). Record the ratio of the number of milliliters of base for each milliliter of acid.

$$\frac{\text{number of ml 2 N NaOH}}{\text{number of ml 1 N HCl}}$$

From this ratio, how many milliliters of your 2 N base neutralize 1 ml of the 1 N acid?

This is called the base equivalent for the 1 N acid. This value is needed for your calculations, as you will note on the table of data.

IV. After test tubes c1 and d1 have been standing about 1 hour (record the exact time), titrate them with your standard NaOH.

V. Test tubes c2 and d2 are titrated about 24 hours later (optional*).

VI. Test tubes c3 and d3 are titrated about 1 week later.

Optional Experiments

VII. Changes in Concentration: Effect on reaction rate and equilibrium constant.

Follow the same method as outlined in the procedure I, but change the concentrations of the reactants as follows:

In test tubes c1, c2, and c3, pipet 5 ml of ethyl acetate, 5 ml of water, 1 ml of 1 N hydrochloric acid.

In test tubes d1, d2, and d3, pipet 8 ml of ethyl alcohol, 2 ml of glacial acetic acid, 1 ml of 1 N hydrochloric acid.

Titrate the resulting solutions as outlined in IV, V, and VI. Record results. Calculate reaction rate and equilibrium constant.

VIII. Changes in Temperature:

In the discussion of this experiment it was noted that only temperature can affect the value of the equilibrium constant. To verify the temperature dependence of the equilibrium constant, repeat the experiment using 25-cm test tubes and maintaining the reacting solutions in a 50°C water bath. Agitate the test tubes frequently to ensure proper heat transfer. Titrate test tubes a after 1/2 hour, test tubes b after 1-1/2 hours, and test tubes c after 2-1/2 hours. By drawing the graph of the results, as

---

* If the titration at 24 hours is not possible, then a titration should be done after 2 weeks. The collected data would then be: tubes c1, d1 after 1 hour; tubes c2, d2, after one week; tubes c3, d3 after 2 weeks.

outlined in the previous procedure, it should be obvious that equilibrium is reached much more quickly than at room temperature. Using the values for test tubes c calculate the equilibrium constant.

i. Is there a difference in the value of K_{eq} at the higher temperature?

Calculations

1. Reaction rate curve.

When all the data have been collected, plot on the same graph paper the number of moles of acetic acid on the y-axis against time on the x-axis for both reactions. The first point for both reactions is the initial concentration of acetic acid before reaction occurs. In your calculations be sure to subtract the amount of sodium hydroxide used to titrate the 1 ml of 1 N hydrochloric acid present as a catalyst (last column of data in the tables).

2. Equilibrium constant.

Use the outline in the data pages to calculate the values of the equilibrium concentrations of the reactants and products. Determine the equilibrium constant for the hydrolysis (from set c3) and for the synthesis (from set d3).

NAME____SECTION____________DATE__________

DETERMINATION OF AN EQUILIBRIUM CONSTANT

DATA

II. Standardization of NaOH

Weight of beaker and NaOH ____________

Weight of beaker ____________

Weight of NaOH ____________

Buret readings (for 20 ml HCl) NaOH solution	Trial I	II	III
Final reading			
Initial reading			
Volume used			
Normality of NaOH			
Normality, average			

Show method of calculation

Determination of the Equilibrium Constants

Table for I, III, IV, V, VI

Test tube	Contents	Date and time	Elapsed hours	ml 2 N NaOH	ml of 2 N base minus base equivalent 1 N HCl
c1 c2 c3	10 ml $CH_3COOC_2H_5$ 5 ml H_2O 1 ml 1 N HCl				
d1 d2 d3	5 ml glacial CH_3COOH 5 ml C_2H_5OH 1 ml 1 N HCl				

Test tube set, Procedure, Part Ic

Buret readings	1 hr. c1	24 hrs. * c2	1 week * c3
Final reading			
Initial reading			
Volume of NaOH solution used			

To determine K_{eq} for the hydrolysis of ethyl acetate, calculate: (c3)

a. Number of moles of acetic acid present, CH_3COOH ________

b. Number of moles of ethyl alcohol present, C_2H_5OH ________

c. Weight of ethyl acetate used (density = 0.901 g/ml) ________

d. Number of moles of ethyl acetate initially present ________

e. Number of moles of ethyl acetate remaining at equilibrium (d - a), $CH_3COOC_2H_5$ ________

f. Number of moles of water initially present (3 ml) ________

g. Number of moles of water at equilibrium, H_2O, (f - a) ________

To obtain experimental values for the equilibrium constants for hydrolysis and synthesis, it is not necessary to express the concentration of all species in moles per liter. The volume is the same for reactants and products. Using moles only will give the same numerical value for the equilibrium constant.

Determine the equilibrium constant for the hydrolysis of ethyl acetate from the above data. Show your calculations.

K_{eq} ____________

* See footnote p. 269.

NAME ______________________________ SECTION ____________ DATE ____________

Test tube set Id

Buret readings	1 hr. d1	24 hrs. * d2	1 week* d3
Final reading			
Initial reading			
Volume of NaOH solution used			

To determine K_{eq} for the formation of ethyl acetate, calculate: (d3)

a. Number of moles of acetic acid at equilibrium (determined by titration with NaOH), CH_3COOH ____________

b. Initial weight of acetic acid (density of CH_3COOH = 1.055) ____________

c. Number of moles of acetic acid initially present ____________

d. Number of moles of ethyl acetate present at equilibrium, $CH_3COOC_2H_5$ (c - a) ____________

e. Initial weight of ethyl alcohol (density = 0.789 g/ml) ____________

f. Number of moles of ethyl alcohol present initially ____________

g. Equilibrium concentration of ethyl alcohol, C_2H_5OH (f - d) ____________

h. Number of moles of water present initially (1 ml) ____________

i. Number of moles of water present at equilibrium, H_2O (h + d) ____________

j. Calculate the equilibrium constant for the formation of ethyl acetate (the reaction of ethyl alcohol and acetic acid) from your experimental data. Show calculations.

K_{eq} = ____________________

VII, VIII Optional Experiments. Using a similar format, calculate K_{eq} for the new conditions.

* See footnote p. 269.

THOUGHT

1. a. You obtained the equilibrium constants for the forward and reverse reaction for the preparation of ethyl acetate. How do these compare?

   b. Which reaction goes further to completion?

2. How are the values for the amount of water in the calculations for set c and set d determined?

3. Given the equation for the hydrolysis of ethyl acetate, what effect if any will each of the following have upon the equilibrium constant?

   a. Increased concentration of water.

   b. Increased concentration of ethyl alcohol.

   c. Increased temperature of reaction.

NAME______________________________________SECTION____________DATE___________

4. What effect will the following have upon the reaction rate of the hydrolysis?

   a. Increased concentration of water.

   b. Increased concentration of ethyl alcohol.

   c. Increased temperature of reaction.

5. In this experiment, how is the course of the reaction followed?

NAME____SECTION____________DATE_____________

Experiment 21
QUALITATIVE ANALYSIS

PRELIMINARY QUESTIONS

1. Define:

   a) decantate

   b) precipitate

2. a) What does the phrase "test for complete precipitation" mean?

   b) Why is it important to test for complete precipitation?

3. a) What are the physical processes symbolized by the diagram?

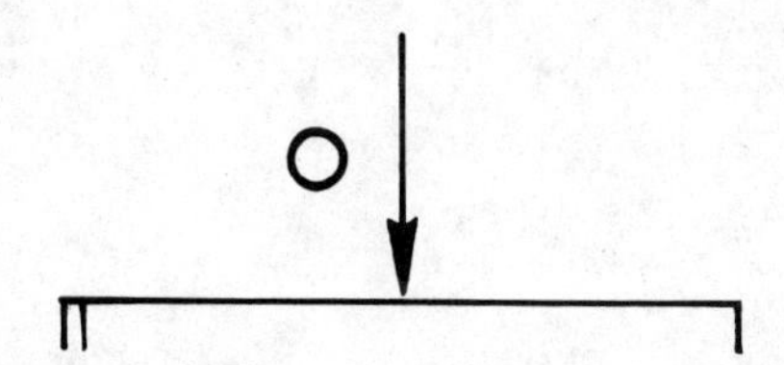

   b) identify the parts of this diagram

4. a) What does it mean to "wash" a precipitate?

   b) Why is it necessary to wash a precipitate?

5. a) What is the rationale for the separation of the cations into groups?

   b) Why is the sequence in which the groups are separated important?

Experiment 21

QUALITATIVE ANALYSIS: CATIONS

IDEAS

Before chemistry matured as a science, attempts to analyze materials consisted of the application of various practical tests that had been collected over thousands of years. For instance, Dioscorides, a Greek physician of the first century A. D., wrote of the test for the adulterate of pompholyx (a ground ore of zinc oxide used for medicinal purposes) by the addition of vinegar which produced a brassy odor, a dark color, and an unpleasant taste. Savonarola, whose grandson was the famous Florentine monk of the fifteenth century, described qualitative tests to differentiate salt from soda. R. Boyle, in the seventeenth century, systematized chemical identification tests, and introduced new ones of his own (as previously mentioned in the discussion of acids and bases). He used silver salts to identify the chloride ion, and sulfuric acid to precipitate calcium salts as the white sulfate. Boyle is considered one of the founders of qualitative analysis, for he was one of the first to use tests consistently to identify substances chemically. The problem of analyzing materials was overwhelming at that time, because it involved the possible addition of impurities with the substances used in analysis; the resulting reactions then could not be clear-cut.

The coherent arrangement of the pieces and bits of information accumulated during centuries was accomplished by K. R. Fresenius (1818-1897), whose orderly scheme of qualitative analysis published in 1841 provided the foundation for modern qualitative analysis. Although the use of this scheme has been superseded by instrumental analysis today, it is only outmoded in its applications. The scheme of qualitative analysis is still theoretically sound, and provides an excellent opportunity for the student to apply many of the fundamental principles of chemistry to solve a problem which is within the scope of his capabilities.

The purpose of this study is to become familiar with the specific properties of selected ions, to apply concepts of aqueous ionic equilibria, and to understand how chemical reactions can be logically ordered to permit the sequential analysis of many ions which may be present in the same solution. Supplemental to this study--often referred to as a "classical wet analysis"--is the application of instrumental and paper chromatographic techniques to the analysis of the ions.

The experience in qualitative analysis is an important introduction to the techniques of observation and analysis in the laboratory. To that end the procedures require tests on individual ions, the recording of these observations and the writing of equations. Once the unique characteristics of the ions are understood, a schematic procedure (flow chart) is presented which utilizes these differences to permit separations and identifications of mixtures of ions. With this schematic it is possible to analyze an "unknown" solution.

The outline of the laboratory procedure is

1. to perform tests on each individual ion and record results on a data page
2. to write net ionic equations for the observations
3. to solve problems which demonstrate the separation of 2 or more ions by chemical reaction
4. to understand the schematic or flow chart
5. to analyze an "unknown" which may contain one or more ions.

This procedure is for the qualitative analysis of 14 cations. The first step is to break down this larger number of ions into several smaller groups consisting of from 3 to 4 ions of similar chemical properties. A separation is always accomplished by using a reagent which forms insoluble precipitates with some but not all of the ions. The precipitate is separated from the solution by centrifuging, then decanting or pipetting the solution above the residue into a second test tube. The analysis is continued on both precipitate and solution each of which now contain fewer ions. The procedure continues until all the ions have been identified.

A flow chart, which is a diagram of the procedure, is included for each of the four cation groups, and also for the procedure below to separate the cations into these groups. Each chart contains a great deal of information, and will be *the guide* to the investigation of the individual cations, and to the solving of the problem of the separation and identification of cations in the groups. Useful notes on the procedure are numbered to coincide with the flow chart.

Techniques for Qualitative Analysis

Successful qualitative analysis is achieved with careful and clean work. Take time to organize.

1. When the directions call for heating a reaction mixture, the test tube is immersed in a hot water bath. The hot water bath is a 150-ml beaker, 3/4 full of water, heated with your bunsen burner.

   Most test tubes have frosted surfaces near the top. These can be marked with a *pencil* for identification of contents. The pencil marks do not wash off or run in the steam, but can be easily erased if you wish to change them.

2. All reactions are stirred thoroughly with a glass *rod* to prevent reactions occuring only at the interface of the two solutions.

3. Whenever water is added to a reaction, use deionized water.

4. Precipitates are packed in the bottom of test tubes by centrifuging. The centrifuge can hold a maximum of six to eight 10-cm test tubes. Since the centrifuge whirls at about 2000 rpm, it must be balanced when in use. Therefore, be sure to pair your test tube with another, filled with solution

or water to the same level, and place the two tubes in opposite slots in the centrifuge.

Run the centrifuge 30-60 seconds, turn off, and slow it gradually by pressing your finger on the top at the center of rotation. Avoid sudden stops which will cause the precipitate to float loose.

5. All precipitates and solutions have to be tested for complete precipitation. To do this after centrifuging and decanting, add a few drops more of the precipitating agent to the solution, and watch for precipitate formation. Add reagent dropwise until no more precipitate forms, stir well, and centrifuge. The test for complete precipitation is abbreviated as T.C.P. in the flow chart.

6. The solution can be separated from the precipitate by
   i) decanting (pouring off) the solution into another test tube. This works well if the precipitate is solidly packed and if you pour carefully.
   ii) by pipetting. Using a medicine dropper with a long thin tip, squeeze the bulb before immersing the tip below the solution level to a point just above the precipitate. Release the bulb slowly, drawing the solution into the pipet. Then transfer the solution to another test tube.

7. For convenience, fill your wash bottle with deionized water. Clean your 400 ml beaker and fill 3/4 full with deionized water. Use the beaker to keep your stirring rods and pipets clean and ready for use.

8. *Preparation of glassware*

   Make the following items:

   3 thin 12-15 cm long glass stirring rods.
   6 glass droppers with long drawn tips; the entire length of the dropper should be about 15 cm of which the tip is about 2 cm.

INVESTIGATION

Chemicals: 0.1M silver nitrate, $AgNO_3$; 0.2M lead nitrate, $Pb(NO_3)_2$; 0.1M mercury(I) nitrate, $Hg_2(NO_3)_2$; 0.1M mercury(II) chloride, $HgCl_2$; 0.1M copper(II) nitrate, $Cu(NO_3)_2$; 0.1M antimony(III) chloride $SbCl_3$; 0.1M cadmium nitrate, $Cd(NO_3)_2$, 0.1M chromium(III) nitrate, $Cr(NO_3)_3$, 0.1M iron(III) nitrate, $Fe(NO_3)_3$, 0.1M aluminum nitrate, $Al(NO_3)_3$, 0.1M nickel nitrate, $Ni(NO_3)_2$, 0.1M barium nitrate, $Ba(NO_3)_2$, 0.1M strontium nitrate, $Sr(NO_3)_2$; 0.1M calcium nitrate, $Ca(NO_3)_2$; 6M hydrochloric acid, HCl; 6M ammonia, NH_3; 6M sodium hydroxide, NaOH; 1M ammonium carbonate, $(NH_4)_2CO_3$; 0.1% methyl violet indicator (aqueous); 1M thioacetamide, CH_3CSNH_2, 0.3M hydrochloric acid, HCl.

Step 1 The Rational For Separation Into Analytical Groups

A. Set up 14 clean 10-cm test tubes in a rack. With *pencil*, write the formula for each cation on the frosted surface of each tube: Ag^+, Hg_2^{2+}, Pb^{2+}, Cu^{2+}, Hg^{2+}, Cd^{2+}, Sb^{3+}, Ni^{2+}, Fe^{3+}, Cr^{3+}, Al^{3+}, Sr^{2+}, Ba^{2+}, Ca^{2+}.

All solutions are 0.1M except Pb^{2+}, which is 0.2M, and all are nitrates except Sb^{3+}, and Hg^{2+} which is present as chloride.

Using a medicine dropper, add 2 drops of a single cation solution to each test tube and 18 drops of deionized* water. (Note that the volume of solution in each test tube is approximately 1 ml.)

To each solution add 5 drops 6M HCl, stir well with a glass rod, centrifuge and note which cation solutions now contain precipitates, what color they are, and whether the precipitate appears crystalline or gelatinous. To those solutions in which precipitates have formed**, add 2 drops more 6M HC Observe closely the solution above the precipitate to see if more precipitate forms. This is a "test for complete precipitation" (T.C.P.) and *must be done* after every test. It ensures that all cations which form the insoluble salts have precipitated and will not interfere by their presence in succeeding procedures.

The *cations* which formed the *insoluble chlorides* are known as *Group I catic* Observe and record their color. Do they have any distinguishing or unique appearance?

B. The solutions of cations which remain are acid. To separate the next grour a careful adjustment of pH to 0.5 (0.3M H^+) is necessary.*** This adjustment is critical. It is done as follows.

Make several spots of methyl violet indicator on a filter paper. Using a prepared solution of 0.3M HCl, add one drop to a methyl violet spot, and observe the blue green color. Test all of the cation solutions in which the precipitation did not occur on the other spots. If the solutions are too acid (the indicator will become more yellow), add NH_3 dropwise to the solution until the color tests match the blue green color obtained with 0.3M HC If the solutions are not acidic enough (purple) add HCl dropwise until the blue-green color of indicator is obtained. Since all the cation solutions were made up the same way, except Sb^{3+}, you can save yourself some time if count the drops of HCl or NH_3 required to obtain the proper pH in the firs cation solution. Then add the identical number of drops of acid or base to

---

* Always use deionized water throughout the analysis.

** If no precipitate has formed, do not continue to add precipitating reagent

*** The purpose of this adjustment is to keep the ions in their proper groups. If the solution is too acidic, some ions from Groups II may not precipitate here but will be carried over to later Groups and interfere with subsequent test. If the solutions are not acidic enough, then some Group III cations will precipitate and interfere with the tests for Group II ions.

of the other solutions. But *you must confirm* the $[H^+]$ for each solution by test.

Once the proper pH has been achieved, add 10 drops thioacetamide* to each solution, stir and heat for 5 minutes. Then centrifuge those solutions which contain precipitates.

Record your observations. Note colors and other distinguishing characteristics. The *cations* which formed the *insoluble sulfides* in *acid solution* belong to *Group II*.

C.** To the cation solutions which remain add 3 drops concentrated HCl and heat (HOOD!) about five minutes. This will remove H_2S from the solutions.*** Now add 6M NaOH until the solution is neutral (test with litmus****) and then add 1 drop more 6M NaOH. Heat. Record the results, (note color and appearance of precipitates.) These *cations* which form *insoluble hydroxides* belong to *Group III*.

D. To the remaining cation solutions add 6M HCl until neutral to litmus, then 5 drops $6M \cdot NH_3$ and 5 drops 1M $(NH_4)_2CO_3$. Record the results. These *insoluble carbonates* are formed by the *Group IV cations*.

THOUGHT

Step I The Rationale for the Group Separations

A. 1. What is the approximate concentration of the cations after dilution with water? Assume 1 drop = 0.05 ml.

2. The cations which belong to Group I are ______________________________.

3. Write net ionic equations for the reactions of the cations with Cl^-.

B. 1. Which cations precipitated with thioacetamide in an acid solution? (Group II cations)

______________________________.

2. What color is each sulfide?

---

* Thioacetamide hydrolyses *slowly* in hot water to yield H_2S.

** If this is not a general unknown, 'C' can be simplified by starting with a new set of cation solutions (those which have not yet been classified into groups). Omit first three lines of Paragraph C, and continue procedure with fourth line "....add 1 drop 6M NaOH. Heat...."

*** Test with moist lead acetate test paper. $Pb(C_2H_3O_2)_2 + H_2S \rightarrow PbS + 2HC_2H_3O_2$. PbS is silvery black.

**** Litmus dye is red in acid solutions and blue in basic solutions.

Summary of cation group separation. Record color and formula of precipitates.

Cation	A. 6M HCl	B. 0.5 pH; H_2S; heat.	C. 6M NaOH	D. 6M NH_3; IM $(NH_4)_2\ CO_3$
Ag^+				
Pb^{2+}				
Hg_2^{2+}				
Hg^{2+}				
Cu^{2+}				
Sb^{3+}				
Cd^{2+}				
Cr^{3+}				
Fe^{3+}				
Al^{3+}				
Ni^{2+}				
Ba^{2+}				
Sr^{2+}				
Ca^{2+}				

3. Write net equations for each reaction. Use the formula H_2S for thioacetamide.*

a.

b.

c.

d.

C. 1. Why does the addition of hydrochloric acid and heat remove H_2S from the solution?

2. Which of the remaining cations precipitate as insoluble hydroxides? (Group III cations)

3. What is the color and formula of each precipitate?

4. Write net ionic equations for the formation of each.

a.

b.

c.

d.

D. 1. Which cations precipitate as insoluble carbonates? (Group IV cations)

---

* $CH_3CSNH_2 + 2H_2O + H^+ \longrightarrow CH_3COOH + NH_4^+ + H_2S$ This reaction occurs in hot water.

2. What is the color and formula of each.

3. Write net equations for the formation of each.

E. Draw a flow chart to show how to separate mixtures of the following cations:

1. Ag^{+}, Hg^{2+}

2. Cu^{2+}, Ni^{2+}

3. Pb^{2+}, Al^{3+}, Ba^{2+}

4. Sr^{2+}, Ni^{2+}

RATIONALE FOR SEPARATION INTO GROUPS
FLOW CHART AND PROCEDURE

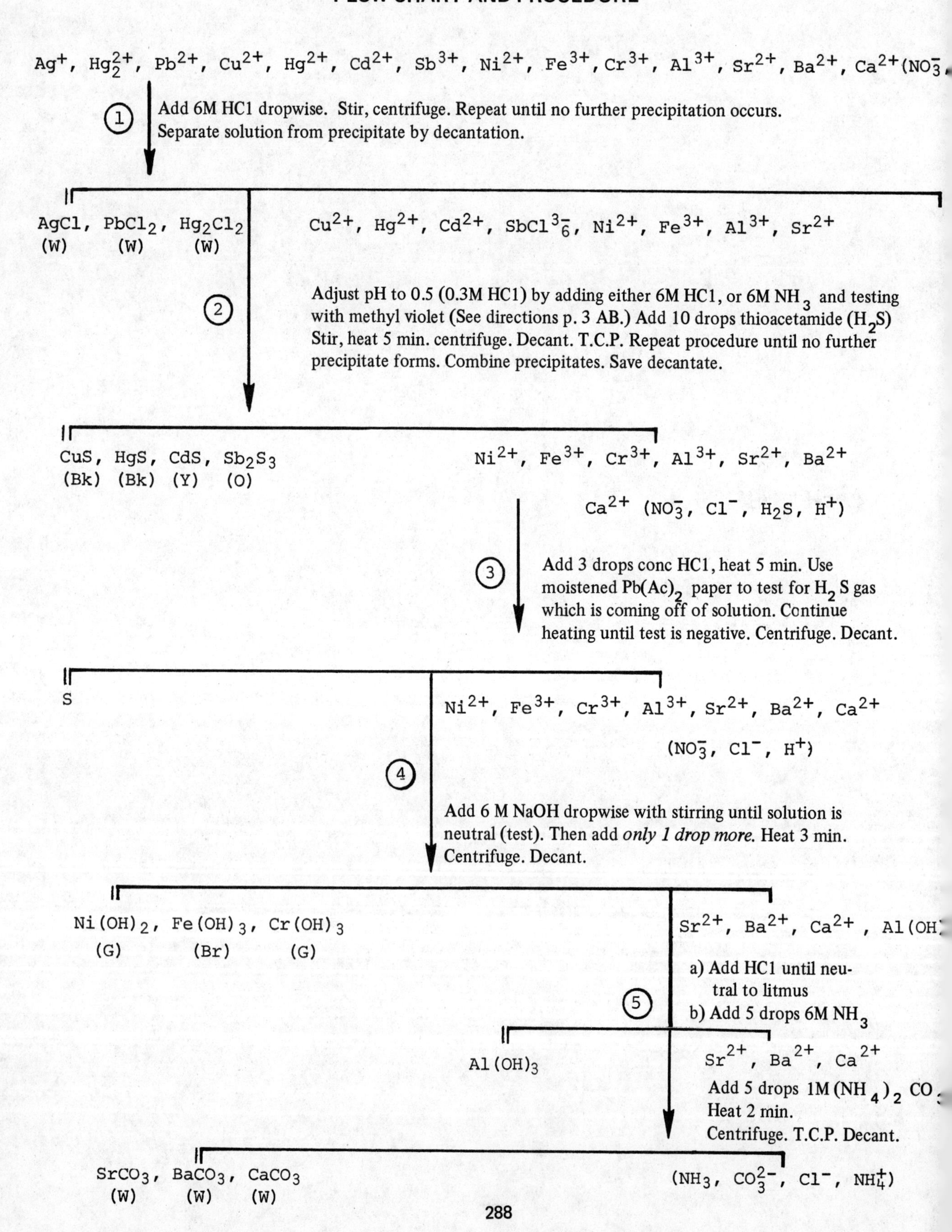

Ag+, Hg2 2+, Pb2+, Cu2+, Hg2+, Cd2+, Sb3+, Ni2+, Fe3+, Cr3+, Al3+, Sr2+, Ba2+, Ca2+ (NO3
1
Add 6M HCl dropwise. Stir, centrifuge. Repeat until no further precipitation occurs.
Separate solution from precipitate by decantation.
AgCl, PbCl2, Hg2Cl2
(W) (W) (W)
Cu2+, Hg2+, Cd2+, SbCl6 3-, Ni2+, Fe3+, Al3+, Sr2+
2
Adjust pH to 0.5 (0.3M HCl) by adding either 6M HCl, or 6M NH3 and testing with methyl violet (See directions p. 3 AB.) Add 10 drops thioacetamide (H2S) Stir, heat 5 min. centrifuge. Decant. T.C.P. Repeat procedure until no further precipitate forms. Combine precipitates. Save decantate.
CuS, HgS, CdS, Sb2S3
(Bk) (Bk) (Y) (O)
Ni2+, Fe3+, Cr3+, Al3+, Sr2+, Ba2+
Ca2+ (NO3-, Cl-, H2S, H+)
3
Add 3 drops conc HCl, heat 5 min. Use moistened Pb(Ac)2 paper to test for H2S gas which is coming off of solution. Continue heating until test is negative. Centrifuge. Decant.
S
Ni2+, Fe3+, Cr3+, Al3+, Sr2+, Ba2+, Ca2+
(NO3-, Cl-, H+)
4
Add 6 M NaOH dropwise with stirring until solution is neutral (test). Then add only 1 drop more. Heat 3 min. Centrifuge. Decant.
Ni(OH)2, Fe(OH)3, Cr(OH)3
(G) (Br) (G)
Sr2+, Ba2+, Ca2+, Al(OH
a) Add HCl until neutral to litmus
5
b) Add 5 drops 6M NH3
Al(OH)3
Sr2+, Ba2+, Ca2+
Add 5 drops 1M (NH4)2 CO
Heat 2 min.
Centrifuge. T.C.P. Decant.
SrCO3, BaCO3, CaCO3
(W) (W) (W)
(NH3, CO3 2-, Cl-, NH4+)

Flow chart diagram interpretation.

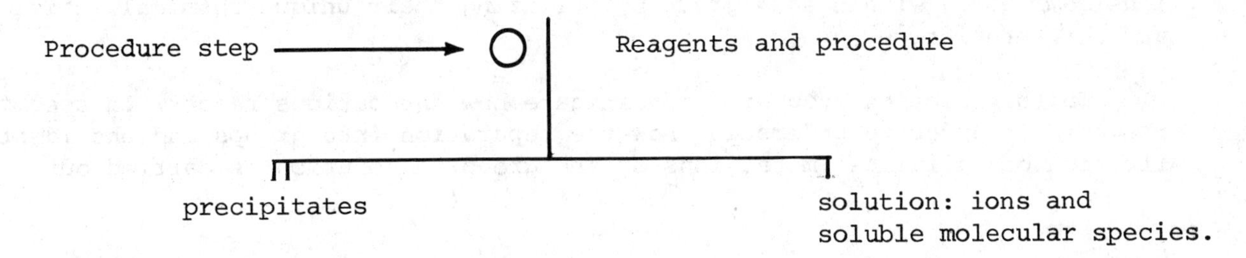

The ┌────┐ indicates the contents of the test tube. These are separated from each other by decantation or pipetting.

The procedure step is coded with the notes and the equations for the reactions. The step, notes, and equations all carry the same number.

Notes. Rationale for Separation into Groups

(1) This removes all cations which form insoluble chlorides. A test for complete precipitation is essential because these cations will interfere in the separation of Groups II-IV by forming insoluble sulfides with thioacetamide.

(2) Control of pH is necessary to allow the precipitation of Group II, but not Group III sulfides. The thioacetamide hydrolyzes to form H_2S in hot water. The hydrolysis is slow, and CdS, the most soluble of the sulfides is slow to form. Therefore, repeated testing for complete precipitation is necessary to prevent these cations from interfering in succeeding tests.

(3) Letters in parenthesis indicate color. For example: W = white, R = red, O = orange, Y = yellow, G = green, Bl = blue, V = violet, Br = brown, Bk = black, and Gy = gray.

(4) Colloidal sulfur may form. Add 5 drops 1 M NH_4NO_3, heat and centrifuge. $Cr(OH)_3$ shows amphoteric properties and will dissolve in excess NaOH, forming $Cr(OH)_4^-$ if the solution is not heated strongly and if too much NaOH is present. $Al(OH)_3$ is also amphoteric. Do not add more than 1 drop 6M NaOH when the solution is neutral.

INVESTIGATION OF GROUP 1 CATIONS
IDEAS

As noted in the outline of laboratory procedure (page 280), the individual cations of each group are studied before an attempt is made to separate and identify the unknown mixture of cations. The presence of individual ions will be demonstrated by the precipitation of characteristically colored insoluble salts, by the formation of a colored solution produced by complexing the cation to a suitable ligand, by the oxidation of the cation to a higher valence state with unique identifying color, or by flame test. For example, Sb^{3+} forms an orange precipitate, Sb_2S_3 with the sulfide ion; Cr^{3+} forms a yellow solution when oxidized to the CrO_4^{2-} ion by sodium peroxide (Na_2O_2), and Fe^{3+} forms a dark red complex ions with SCN: $FeSCN^{2+}$.

These confirmatory tests, however, must be free of interference by ions which may be present in the solution. It is therefore necessary to separate the ions *completely* within each group by utilizing their unique chemical activity and characteristics.

In this section, you will investigate how the cations react with specific reagents in order to understand how the separation into groups and the identification of individual metal ions of the group in solution is carried out.

INVESTIGATION

CHEMICALS: 0.1M silver nitrate, $AgNO_3$: 0.1M mercury (I) nitrate, $Hg_2(NO_3)_2$; 0.2M lead nitrate, $Pb(NO_3)_2$; 6M hydrochloric acid, HCl; 6M ammonia, NH_3; 1.0 M potassium chromate, K_2CrO_4; 6M nitric acid, HNO_3.

A. Put 2 drops of 0.1M Ag NO_3, 0.2M $Pb(NO_3)_2$, and 0.1M $Hg_2(NO_3)_2$ and 18 drops of water in 3 separate test tubes. Add 4 drops 6M HCl to each tube and observe the results -- i.e., precipitate formation, appearance of precipitate, texture and color, and change in solution appearance. Record the observations in Table I, page 291. Heat each of the test tubes in a boiling water bath, and record results in the table. Cool the tubes by immersing them in cold water, then centrifuge. Decant the solution from the test tubes which contain the Hg_2Cl_2 and AgCl precipitates; discard the decantate. Add 1 ml deionized water to the precipitates, stir well, centrifuge, and discard the wash water. Add 5 drops 6M NH_3 to each, stir well. Solution must be basic. Test. Add more NH_3 if necessary. Record your results.

B. To the test tube containing the Ag^+ ion, add 6 drops 6M HNO_3, test to be sure the solution is acid, and record the results.

C. To 3 new samples containing 2 drops each 0.1M $AgNO_3$, 0.1M $Hg_2(NO_3)_2$ and 0.2M $Pb(NO_3)_2$ in separate test tubes and 18 drops of water, add 5 drops 1.0M K_2CrO_4 and record the results. This reaction is a confirmatory test for Pb^{2+} ions. Do you see why?

Now proceed to do an unknown analysis. Follow the instructions in the flow chart p. 294.

Table I

Characteristics of Group I Cations

Record formula of products as well as precipitate formation, color changes, or other unique appearances.

Reagent	Ag^+	Pb^{2+}	Hg_2^{2+}
6M HCl			
Heat			
6M NH_3		X	
6M HNO_3		X	X
1.0M K_2CrO_4			

THOUGHT

1. Complete the net ionic equations for the following:
   a. Hydrochloric acid with

      Ag^+

      Pb^{2+}

      $Hg_2{}^{2+}$

   b. Heat the products from (a) with water.

   c. Ammonia with

      AgCl

      Hg_2Cl_2

   d. Nitric acid with products from (c).

e. Potassium chromate with

Pb^{2+}

2. a. Draw a flow chart showing the simplest way to separate Pb^{2+} from Ag^{+}.

b. Draw a flow chart showing the simplest way to separate Ag^{+} from Hg_2^{2+}.

c. If a solution contained only one cation from Group I (Ag^+, Hg_2^{2+}, Pb^{2+}) how would you identify which cation it was? Describe your method as briefly as possible.

GROUP I CATIONS
FLOW CHART AND PROCEDURE

Solutions may contain one, two or three Group I cations, as nitrates, in concentrations of 0.01M. (Pb^{++} is 0.02M) If this is a general unknown, the cations of Groups II-IV may also be present.

Use 2 ml of the unknown solution *without further dilution* for the analysis.

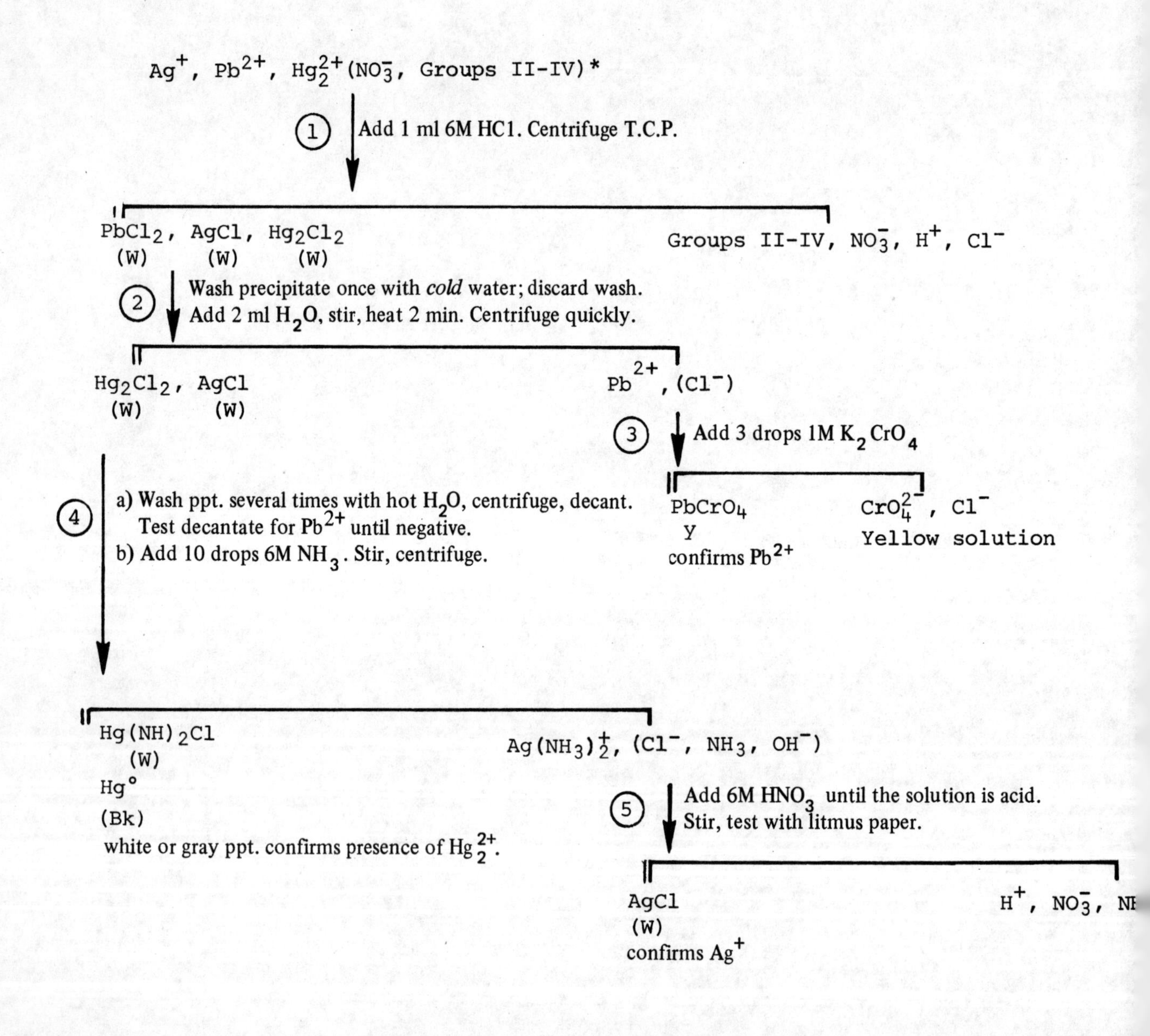

Ag+, Pb2+, Hg2 2+ (NO3-, Groups II-IV)*
1 Add 1 ml 6M HCl. Centrifuge T.C.P.
PbCl2, AgCl, Hg2Cl2
(W) (W) (W)
Groups II-IV, NO3-, H+, Cl-
2 Wash precipitate once with cold water; discard wash.
Add 2 ml H2O, stir, heat 2 min. Centrifuge quickly.
Hg2Cl2, AgCl
(W) (W)
Pb2+, (Cl-)
3 Add 3 drops 1M K2CrO4
PbCrO4
Y
confirms Pb2+
CrO4 2-, Cl-
Yellow solution
4 a) Wash ppt. several times with hot H2O, centrifuge, decant.
Test decantate for Pb2+ until negative.
b) Add 10 drops 6M NH3. Stir, centrifuge.
Hg(NH)2Cl
(W)
Hg°
(Bk)
white or gray ppt. confirms presence of Hg2 2+.
Ag(NH3)2+, (Cl-, NH3, OH-)
5 Add 6M HNO3 until the solution is acid.
Stir, test with litmus paper.
AgCl
(W)
confirms Ag+
H+, NO3-, NH

My unknown contained the following cations ______________________

*See page 289 for an explanation of the symbols.

Notes. Group I Cations

① Parentheses enclose ions which are present because they have been added as reagents, are anions of the salts, or are cations of succeeding groups.

T.C.P. means "test for complete precipitation." Here in Step 1 it means add 2 more drops 6M HCl and observe the supernatant solution carefully for the formation of more precipitate. Stir the solution, heat one minute, cool and centrifuge. This process continues until no precipitates form when 6M HCl is added.

② Avoid allowing the solution to cool. $PbCl_2$ will precipitate at room temperature.

③ If a precipitate forms here, the color of it can be readily seen if the yellow solution is decanted, followed by the addition of several drops of deionized H_2O to the precipitate. Stir well and centrifuge.

If the solution to which the K_2CrO_4 has been added is orange, it is too acid. Add 1-2 drops NH_3. *Yellow* precipitate indicates $PbCrO_4$.

ppt. = abreviation for precipitate.
sol'n = abreviation for solution.

④ If lead is present, care must be taken to remove it completely from this ppt. Repeated hot water washings of the AgCl, Hg_2Cl_2 ppts. are necessary. The Pb^{2+} has been completely removed when the decantate no longer gives a positive yellow ppt. of $PbCrO_4$.

⑤ The silver ammonia complex ion is pH dependent. Hence, care must be taken to be sure this solution is acid. Test with litmus paper

INVESTIGATION OF GROUP II CATIONS
IDEAS

These cations form insoluble sulfides in acid solutions according to the general equation $M^{2+} + H_2S \longrightarrow MS + 2H^+$. ($M^{2+}$ represents any metallic positively charged ion. In this scheme the charge is either 2^+ or 3^+.) The directions involve careful adjustment of the pH of the solution to 0.5. The purpose is simply to assure that cations of Group III, which form insoluble sulfides at a higher pH, remain in solution. Group III will be precipitated as insoluble hydroxides, rather than sulfides.

The H_2S is produced by the hydrolysis of thioacetamide in *warm* aqueous solutions.

$CH_3CSNH_2 + 2H_2O \rightleftharpoons H_2S + CH_3COO^- + NH_4^+$ The reagent is a 1M solution and dispensed dropwise. In this scheme, thioacetamide is called H_2S.

INVESTIGATION

Chemicals: 0.1M copper nitrate, $Cu(NO_3)_2$; 0.1M cadmium nitrate, $Cd(NO_3)_2$; 0.1M mercury(II) chloride, $HgCl_2$; 0.1M antimony(III) chloride, $SbCl_3$; 1M thioacetamide, CH_3CSNH_2; IM ammonium nitrate, NH_4NO_3; 6M sodium hydroxide, NaOH; 6M nitric acid, HNO_3; 6M ammonia, NH_3 6M hydrochloric acid, HCl; 12M hydrochloric acid, HCl; copper wire or penny, Cu.

1. a. Put 2 drops 0.1M $Cu(NO_3)_2$, 0.1M $Cd(NO_3)_2$, 0.1M $HgCl_2$ and 0.1M $SbCl_3$ in separate labeled 10 cm test tubes. Add 18 drops H_2O and adjust pH to 0.5 as described on p. 282. Add 10 drops thioacetamide (H_2S) stir well and warm gently in a boiling water bath for 5 minutes. Centrifuge and record results. Note the color* of each precipitate and write the formula for each in Table IIA on page 297.

   b. Decant the supernatant liquid, wash the precipitates in each test tube with 10 drops H_2o deionized water containing 1 drop 1M NH_4NO_3, centrifuge, and discard the wash. *This is important.* If excess sulfide ion is not removed, a soluble mercury sulfide complex will form in basic medium: $HgS + S^{2-} \longrightarrow HgS_2^{2-}$)

2. Add 5 drops 6M NaOH, stir, heat. Do any precipitates dissolve?

3. a. To the test tubes containing precipitates add 1 ml deionized water, stir centrifuge, and discard wash. Add 10 drops 6M HNO_3, stir, and warm for 3 minutes. Do any precipitates dissolve? Which do not dissolve?

   b. Add enough 6M NH_3 to each solution to make it basic to litmus. Record your observations, then add an additional 10 drops of 6M NH_3 and record your observations on Table IIA on page 297.

4. Reactions of cadmium

   To the test tubes containing the $Cu(NH_3)_4^{2+}$ and the $Cd(NH_3)_4^{2+}$ add 10 drops 6M NaOH, stir well, and heat 1 minute. If a precipitate forms, centrifuge and decant the supernatant liquid. Wash the precipitate with 5 drops deionized water and discard the wash water. To the precipitate, add 5 drops water, then add 6M HCl dropwise until the pH is adjusted to 0.5. Then add 5 drops thioacetamide. A yellow precipitate of CdS should form. Record your observations on Table IIB on page 297.

---

* HgS has 2 allotropic forms; one is red and the other is black. The allotropic forms depend upon S^{2-} concentration.

Table IIA

Characteristics of Group II Cations

Record formulas of products as well as appearances.

Reagents	Cu^{2+}	Cd^{2+}	Hg^{2+}	Sb^{3+}
1a. H_2S in 0.5 pH				
2. Add 6M NaOH to results above				
3a. Acidify the solutions in (2) with 6M HNO_3				X
3b. Limited NH_3			X	X
3c. Excess NH_3			X	X

Table IIB

Characteristics of Group II Cations (cont'd)

Record formulas and appearances of products.

	$Cu(NH_3)_4^{2+}$	$Cd(NH_3)_4^{2+}$
4. Add 6M NaOH		
Add 6M HCl to ph 0.5 add H_2S	X	

THOUGHT

Draw a flow chart showing the simplest way to separate the following pairs of cations.

1. Sb^{3+} and Cd^{2+}

2. Cu^{2+} and Hg^{2+}

3. Cu^{2+} and Cd^{2+}

GROUP II CATIONS
FLOW CHART AND PROCEDURE

Solutions may contain 0.1M Cu^{2+}, 0.1M Hg^{2+}, 0.1M Cd^{2+}, 0.1M Sb^{3+} as NO_3^- or Cl^-, diluted 10 fold with deionized water. Use 2 ml.

Cu^{2+}, Hg^{2+}, Cd^{2+}, Sb^{3+} (NO_3^-, Cl^-, Groups III, IV)

Step (1):

a) Adjust pH to 0.5 with methyl violet by adding 6M HC1 or 6M NH_3.

b) Add 10 drops thioacetamide* (H_2S), heat 5 min., centrifuge.

c) Decant. TCP decantate. Combine ppt's. Save decantate for Groups III, IV if present.

Precipitate: CuS, HgS, CdS, Sb_2S_3
(BK) (Bk) (Y) (O)

Decantate: NO_3^-, Cl^-, Groups III, IV, H_2S

Step (2):

a) Wash the precipitate with 10 drops water containing 1 drop H_2S. Centrifuge. Discard wash.

b) Add 8 drops 6M NaOH to the ppt., stir, heat 3 min. Centrifuge. Decant. Repeat with 2 drops 6M NaOH. Centrifuge. Combine decantates.

Precipitate: CuS, HgS, Cds
(Go to Step (3) on page 302.)

Decantate: $Sb(OH)_4^-$, SbS_2^-, [HgS_2^{2-}] **
(Go to step (9) on page 304.)

* Thioacetamide will be called H_2S in the procedure from here on.

** Square brackets are used in this outline to indicate a product not expected to form, but which will do so under some conditions.

Notes. Group II Cations

(1) a) Place several drops of methyl violet on a filter paper. To one spot add 1 drop 0.3M HCl (0.5pH) and observe the blue-green color. Adjust the acidity of your cation solution to this pH by adding 6M HCl dropwise if the solution is too basic (spot will be purple) or 6M NH_3 if the solution is too acid (spot will be yellow).

The pH adjustment is crucial: if the solution is too basic, some cations from Group III may precipitate (NiS); if too acid, CdS and Sb_2S_3 may not precipitate.

b) Thioacetamide hydrolyzes slowly in hot H_2O to yield H_2S.

$$CH_3CSNH_2 + 2H_2O \longrightarrow CH_3COO^- + NH_4^+ + H_2S$$

(2) a) A colloidal dispersion may form. Add 5 drops 1M NH_4NO_3 to the solution. Stir, heat, centrifuge. Discard wash.

HgS_2^{2-}, a soluble complex sulfide ion of mercury may form if the S^{2-} concentration is high enough. Its presence will be detected in Step 10, as a black sulfide precipitate. Sb_2S_3 is an acid sulfide and will react with NaOH to form the soluble complex hydroxide. It too can form a soluble sulfide complex if the S^{2-} concentration is high enough.

GROUP II CATIONS
FLOW CHART AND PROCEDURE (continued)

Analysis of the precipitate from Step 2

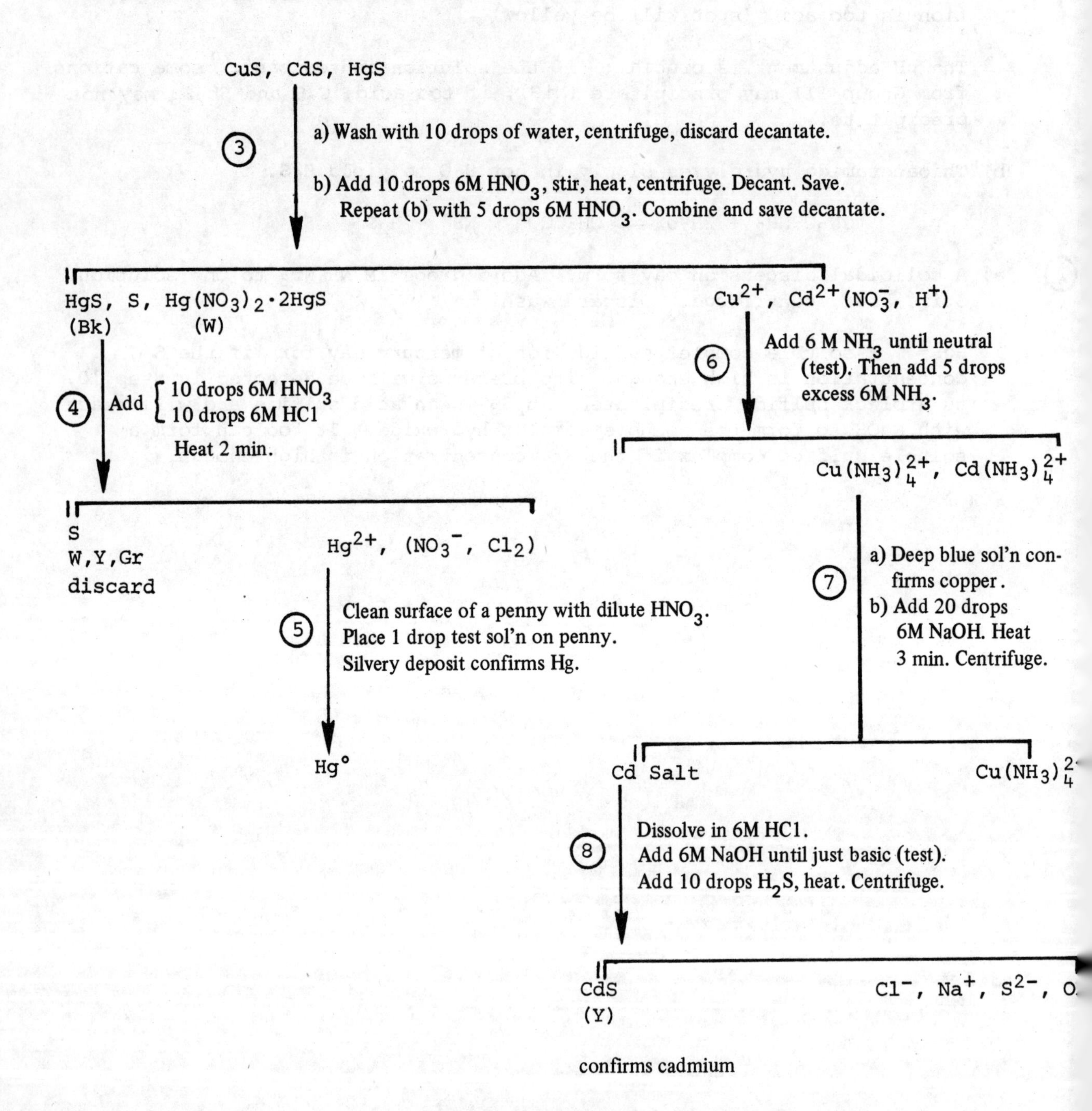

CuS, CdS, HgS
3
a) Wash with 10 drops of water, centrifuge, discard decantate.
b) Add 10 drops 6M HNO3, stir, heat, centrifuge. Decant. Save.
Repeat (b) with 5 drops 6M HNO3. Combine and save decantate.
HgS, S, Hg(NO3)2·2HgS
(Bk) (W)
Cu2+, Cd2+ (NO3-, H+)
6
Add 6 M NH3 until neutral (test). Then add 5 drops excess 6M NH3.
4
Add
10 drops 6M HNO3
10 drops 6M HCl
Heat 2 min.
Cu(NH3)4 2+, Cd(NH3)4 2+
S
W,Y,Gr
discard
Hg2+, (NO3-, Cl2)
7
a) Deep blue sol'n confirms copper.
b) Add 20 drops 6M NaOH. Heat 3 min. Centrifuge.
5
Clean surface of a penny with dilute HNO3.
Place 1 drop test sol'n on penny.
Silvery deposit confirms Hg.
Hg°
Cd Salt
Cu(NH3)4 2
8
Dissolve in 6M HCl.
Add 6M NaOH until just basic (test).
Add 10 drops H2S, heat. Centrifuge.
CdS
(Y)
Cl-, Na+, S2-, O
confirms cadmium

Notes

(3) HgS is extremely insoluble, and will not dissolve with HNO_3. A white, gray, black or red precipitate may indicate mercury.

(4) HgS dissolves in a combination of acids, HNO_3 and HCl, known as aqua regia. HNO_3 oxidizes the S^{2-} to S^0; the Cl^- complexes the Hg^{2+} to form $HgCl_4^{2-}$.

$$3HgS + 2NO_3 + 8H^+ + 12Cl^- \longrightarrow 3HgCl_4^{2-} + 2NO + 3S + 4H_2O$$

(6) a) It is essential that the solution be distinctly ammoniacal. The ammonia complexes are pH dependent and will form only in strongly ammoniacal solutions. An insoluble precipitate $Cd(OH)_2$ may form here in addition to the $Cd(NH_3)_4^{++}$.

(7) b) $NaOH$ and heat will cause the formation of an insoluble cadmium salt. $Cu(NH_3)_4^{2+}$ will remain in solution.

(8) CdS may be slow in forming. There may be a dark rather than yellow ppt. due to contamination by CuS.

GROUP II CATIONS
FLOW CHART AND PROCEDURE (continued)

Analysis of the solution from Step 2

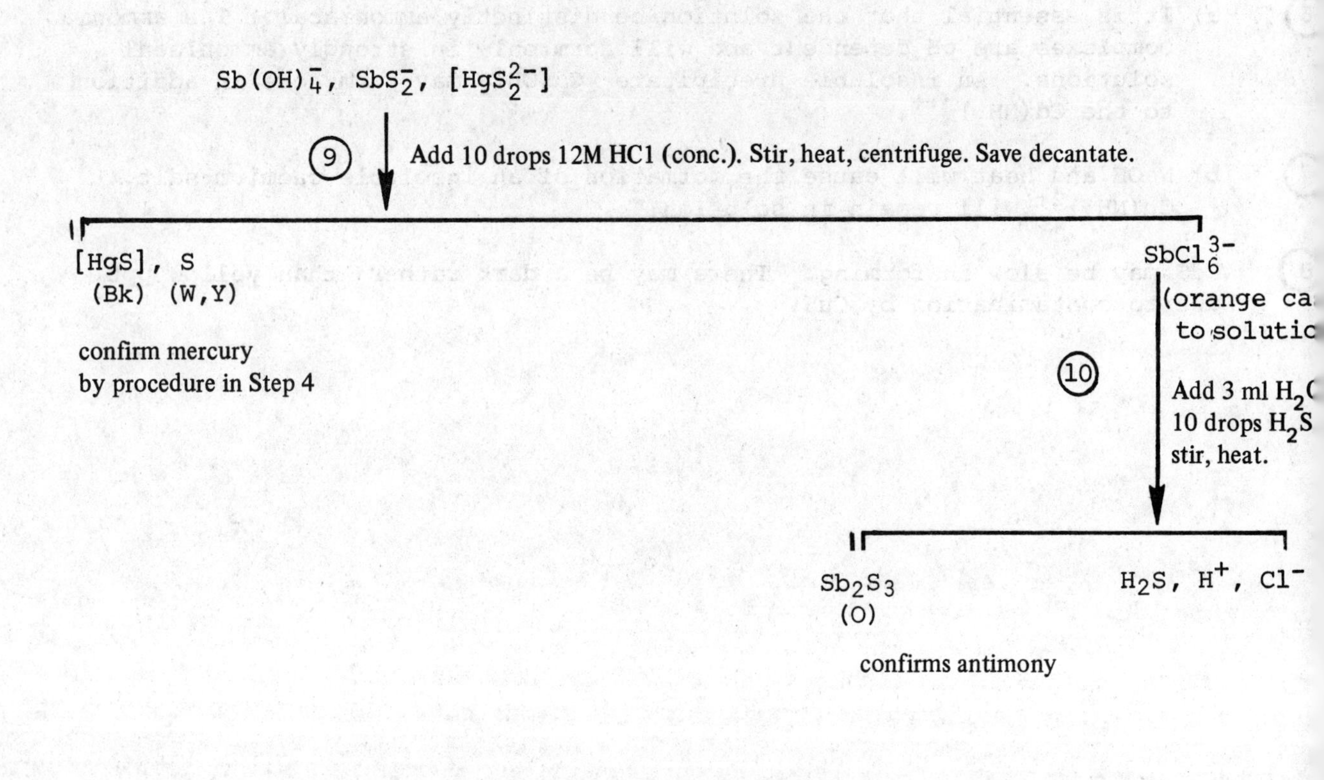

Sb(OH)4-, SbS2-, [HgS2 2-]
9
Add 10 drops 12M HCl (conc.). Stir, heat, centrifuge. Save decantate.
[HgS], S
(Bk) (W,Y)
confirm mercury
by procedure in Step 4
SbCl6 3-
(orange ca
to solutic
10
Add 3 ml H2(
10 drops H2S
stir, heat.
Sb2S3
(O)
H2S, H+, Cl-
confirms antimony

My unknown contained the following cations:______________________________

Notes

(9) An orange precipitate here is probably antimony sulfide. If it does not readily dissolve in the 12M HCl, add 3 drops more acid, stir and heat again.

(10) To precipitate Sb_2S_3, the acidity of the solution must be reduced. If no precipitate forms, and antimony is believed to be present, add 1 drop 6M NaOH, 5 drops H_2S, stir, heat, centrifuge.

GROUP III CATION ANALYSIS
IDEAS

Group III cation precipitation depends upon the formation of insoluble hydroxides. However, some of these cations form soluble complexes with sodium hydroxide or ammonia so it is very important to add exactly the right amount of reagent to form the desired products. For example, aluminum hydroxide will dissolve in excess NaOH, but not in excess buffered NH_3. Nickel ion will form complex ammines in excess NH_3, but with NaOH will form insoluble hydroxides which do not dissolve in excess base. Chromium hydroxide which can form a soluble complex in excess strong base, will not do so if it is in a hot solution.

The separation of cations in Group III, therefore, is subtle. While performing the tests on the individual ions, pay very close attention to the specific conditions which allow precipitation formation and precipitate solution due to the presence of excess reagent.

INVESTIGATION

Chemicals: 0.1M aluminum nitrate, $Al(NO_3)_3$; 0.1M chromium nitrate, $Cr(NO_3)_3$; 0.1M iron(III) nitrate, $Fe(NO_3)_3$, 0.1M nickel nitrate, $Ni(NO_3)_2$; 6M hydrochloric acid, HCl; 6M sodium hydroxide, NaOH; 6M ammonia, NH_3; 0.1M potassium thiocyanate, KSCN; 1.0M barium chloride, $BaCl_2$; 1% dimethylglyoxime, $C_4H_8N_2O_2$, in 95% ethanol; sodium peroxide, Na_2O_2, solid; aluminon reagent, ammonium salt of aurin tricarboxylic acid, 0.1% ; hydrogen peroxide, 3% H_2O_2; 1.0M sodium hydroxide, NaOH.

Perform each test in a 10-cm test tube using 2 drops of 0.1M cation solution diluted with 18 drops of water. The volume of solution should be about 1 ml.

1. Prepare the following properly diluted cation solutions (see above), $Al(NO_3)_3$, $Cr(NO_3)_3$, $Fe(NO_3)_3$ and $Ni(NO_3)_2$, in separate 10-cm test tubes.

   a. Add 3 drops 1M HCl. This is added to prevent an initial excess of OH^- ions.

   Add 6M NaOH dropwise until ppt. appears. Stir, heat. Record your observations in the table.

   b. Add an additional 3 drops 6M NaOH, stir, and heat. Record your observations in the table.

2. a. To a new set of cation solutions, add 3 drops 6M HCl. Then add 6M NH_3 dropwise until a precipitate appears. (A buffer solution -- NH_3-NH_4^+-- forms.)

   b. Add an additional 10 drops of 6M NH_3. Stir and record your observations.

3. To a new set of cation solutions, add 3 drops 0.1M KSCN, stir well and record your observations.

4. To a new set of cation solutions, add 5 drops dimethylglyoxime. Stir well

and record your observations.

5. To a diluted solution of the Cr^{3+} ion, add 6M NH_3 until the solution is basic. Then add a few grains of Na_2O_2 slowly with stirring; heat gently until the solution becomes distinctly yellow in color. Add 5 drops 0.1M $BaCl_2$, centrifuge, T.C.P., and discard supernatant liquid.

   Add 6M HCl to the precipitate dropwise with stirring after each addition until the precipitate just dissolves. (If too much Na_2O_2 or HCl are added, the solution may turn blue at this point indicating the presence of CrO_5. The blue color will fade rapidly to form a green solution due to the presence of the Cr^{3+} ion.) When the precipitate has completely dissolved, the solution should be orange. Place 2 drops of the 3% H_2O_2 solution on a filter paper followed by one drop of the cation solution directly on top of the peroxide. A blue color (CrO_5) which fades rapidly is indicative of chromium.

Table III

Characteristics of Group III Cations

Record formulas of products as well as appearance

Reagent	Al^{3+}	Cr^{3+}	Fe^{3+}	Ni^{2+}
1a. 6M HCl 1M NaOH (limited)				
1b. 6M NaOH (excess)				
2a. 6M HCl 6M NH_3 (limited) (Buffer sol'n NH_3-NH_4^+ Forms)				
2b. 6M NH_3 (excess)				
3. 0.1M KSCN				
4. 1% dimethyl- glyoxime in alcohol				
5. 6M NH_3 Na_2O_2 1.0M $BaCl_2$	X		X	X

1. Using a flow chart, show the simplest way to separate the following pairs.

a. Fe^{3+} from Al^{3+}

b. Ni^{2+} from Cr^{3+}

c. Ni^{2+} from Fe^{3+}

GROUP III CATIONS
FLOW CHART AND PROCEDURE

Solutions may contain 0.1M cations as NO_3^- which have been diluted 10 fold with deionized water. Use 2 ml except for Step (1). If this is not a general unknown, do steps (1) and (3). Omit (2).

Preliminary test for ion.

Fe^{3+}, Ni^{2+}, Al^{3+}, Cr^{3+} (Cl^-, NO_3^-, H_2S, Group IV cations)

(1) Add 2 drops 0.1M KSCN to ½ ml of solution.

I: $FeSCN^{2+}$ — red sol'n confirms iron

Ni^{2+}, Al^{3+}, Cr^{3+} (Group IV cations)

Step (2) Do this only if this analysis is continued from Group II. Otherwise omit (2), and continue the analysis at step (3) with a fresh sample of your unknown.

Fe^{3+}, Ni^{2+}, Al^{3+}, Cr^{3+}, (Cl^-, NO_3^-, H_2S, Group IV cations)

(2) Transfer your solution to a 50-ml beaker.
Add 5 drops concentrated HC1 (12M) and heat until all H_2S is evolved. Test with lead acetate paper. When the test is negative, return the solution to a 10-cm test tube.

II

Fe^{3+}, Ni^{2+}, Al^{3+}, Cr^{3+}, (Cl^-, NO_3^-, Group IV cations)

(3) Add 6M NaOH until basic (test).

Add 1 ml excess 6M NaOH, stir, heat 3 min. T.C.P.
Decant. Save decantate.

II: $Fe(OH)_3$ (R), $Ni(OH)_2$ (G), $Cr(OH)_3$ (Gy)

(Go to Step (4) p.312)

$Al(OH)_4^-$, $Cr(OH)_4^-$

(OH^-, Na^+), Group IV cations

(Go to Step (11) p.314)

Notes. Group III Cations

(1) The test for Fe^{3+} is done on the Group III solution directly. The test is so sensitive that the presence of iron can be detected at very low concentration even if the other cations are present. It will not be necessary to confirm iron in any subsequent procedure.

(2) The separation and identification of the remaining ions begins here when the analysis is continued from Group II.

a) The addition of concentrated HCl shifts the equilibrium of H_2S in aqueous media, toward the molecular form, and the hot solution makes the gas less soluble.

$$2H^+ + S^{2-} \longrightarrow H_2S\uparrow$$

b) All S^{2-} ion must be removed from the solution so that insoluble sulfides of FeS, NiS do not precipitate. To determine when all H_2S has been removed, hold a moistened strip of lead acetate test paper over the mouth of the hot test tube. A blackening or silvering of the paper indicates the presence of H_2S because of the reaction of S^{2-} with the lead acetate to form lead sulfide.

$$Pb(C_2H_3O_2)_2 + H_2S \longrightarrow \underset{(Bk)}{PbS} + 2HC_2H_3O_2$$

(3) $Cr(OH)_3$ does not show amphoteric properties if the solution is heated several minutes in boiling water.

GROUP III CATIONS

FLOW CHART AND PROCEDURE (continued)

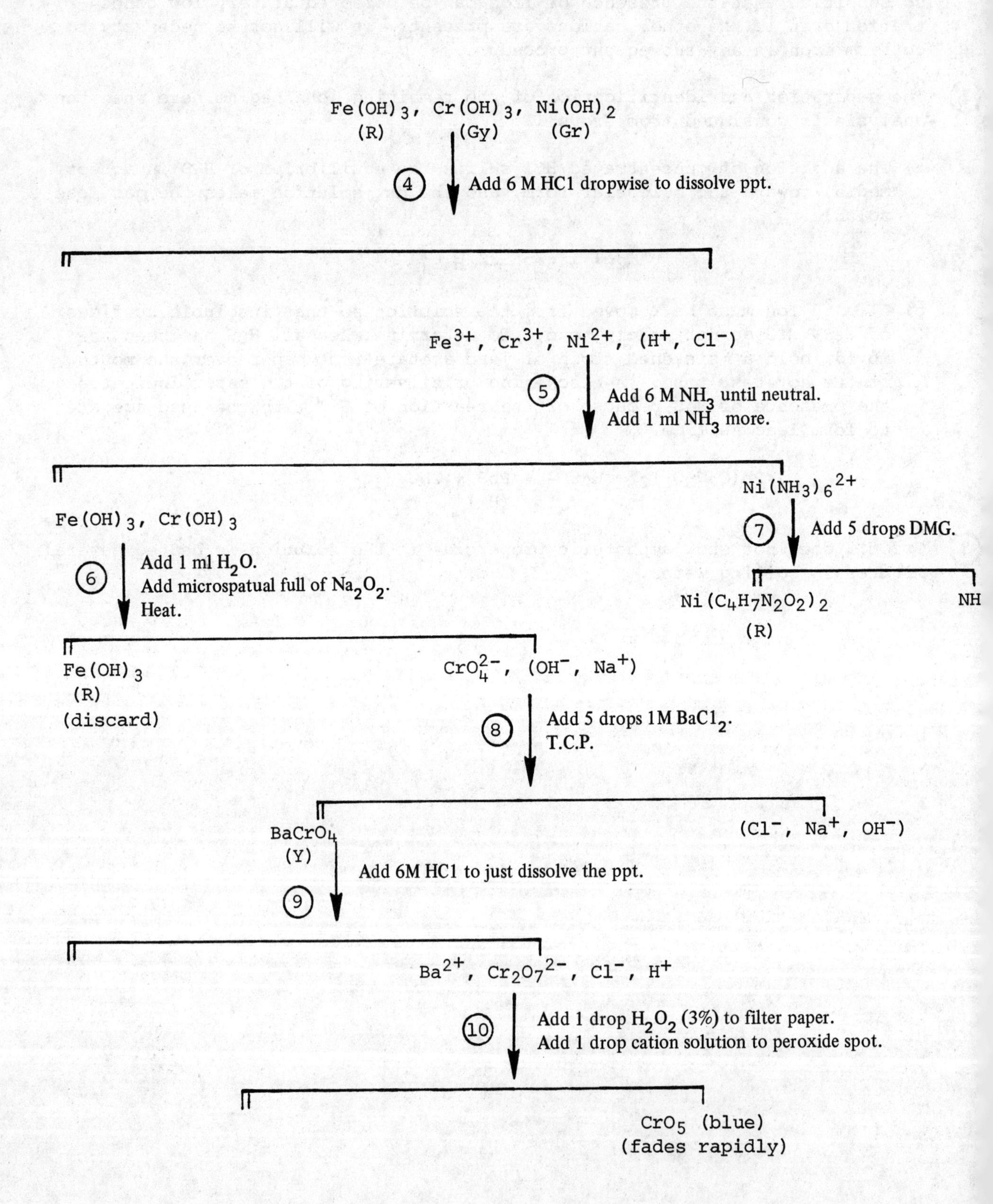

Fe(OH)3, Cr(OH)3, Ni(OH)2
(R) (Gy) (Gr)
4
Add 6 M HC1 dropwise to dissolve ppt.
Fe3+, Cr3+, Ni2+, (H+, Cl-)
5
Add 6 M NH3 until neutral.
Add 1 ml NH3 more.
Fe(OH)3, Cr(OH)3
Ni(NH3)6 2+
6
Add 1 ml H2O.
Add microspatual full of Na2O2.
Heat.
7
Add 5 drops DMG.
Ni(C4H7N2O2)2
(R)
NH
Fe(OH)3
(R)
(discard)
CrO4 2-, (OH-, Na+)
8
Add 5 drops 1M BaC12.
T.C.P.
BaCrO4
(Y)
(Cl-, Na+, OH-)
9
Add 6M HC1 to just dissolve the ppt.
Ba2+, Cr2O7 2-, Cl-, H+
10
Add 1 drop H2O2 (3%) to filter paper.
Add 1 drop cation solution to peroxide spot.
CrO5 (blue)
(fades rapidly)

Notes

(6) A microspatula holds approximately 0.01 g. It is not necessary to weigh.

(8) A yellow solution at this point indicates the presence of chromium (CrO_4^{2-}).

(9), (10) Reduction of the $Cr_2O_7^{2-}$ can occur if either too much acid or too much peroxide is used. Be sure to add only enough HCl to just dissolve the precipitate, $BaCrO_4$.

$$Cr_2O_7^{2-} + 3H_2O_2 + 8H^+ \longrightarrow 2Cr^{3+} + 3O_2 + 7H_2O$$

The reaction to produce the CrO_5 is

$$Cr_2O_7^{2-} + 4H_2O_2 + 2H^+ \longrightarrow 2CrO_5 + 5H_2O$$

The CrO_5 fades in excess peroxide as follows:

$$2CrO_5 + 6H^+ + 7H_2O_2 \longrightarrow 2Cr^{3+} + 10H_2O + 7O_2$$

GROUP III CATIONS
FLOW CHART AND PROCEDURE (continued)

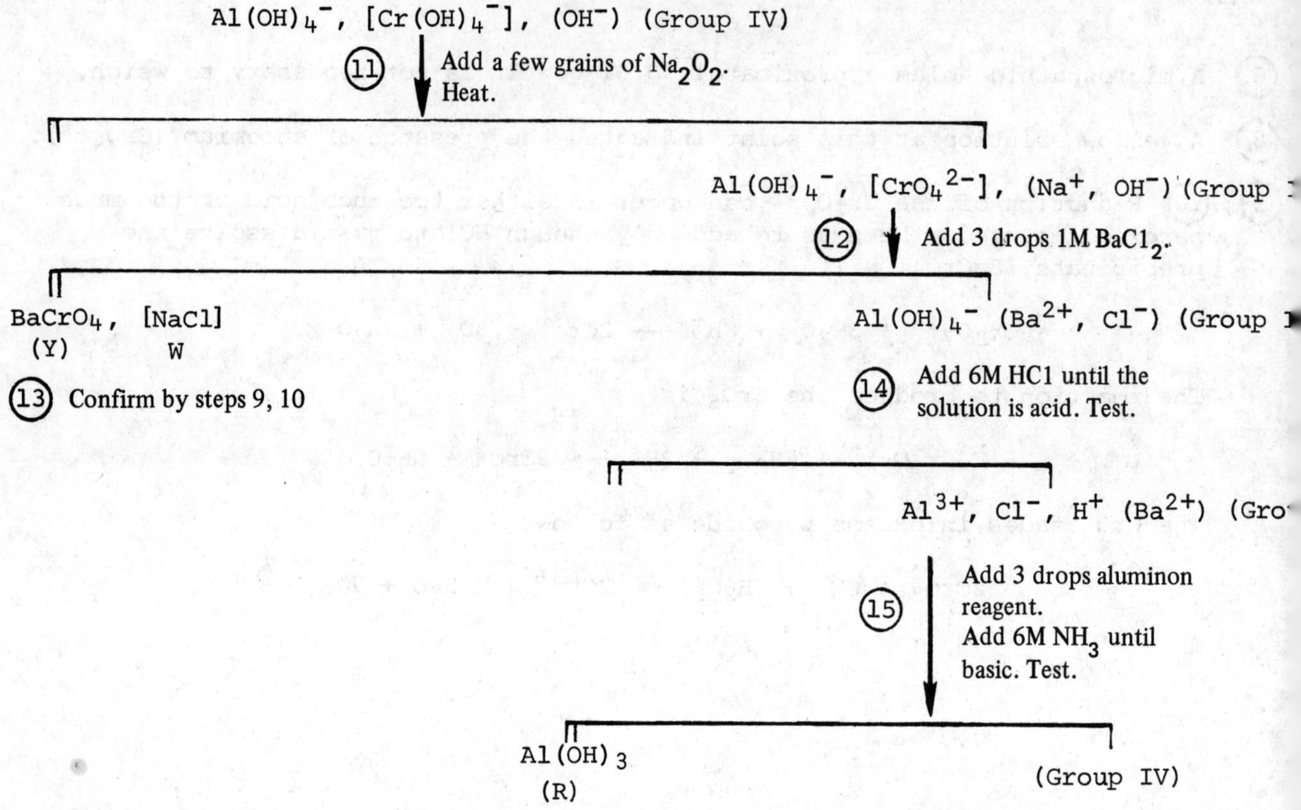

$Al(OH)_4^-$, [$Cr(OH)_4^-$], (OH^-) (Group IV)

(11) Add a few grains of Na_2O_2. Heat.

$Al(OH)_4^-$, [CrO_4^{2-}], (Na^+, OH^-) (Group I

(12) Add 3 drops 1M $BaCl_2$.

$BaCrO_4$, [NaCl]
(Y) W

(13) Confirm by steps 9, 10

$Al(OH)_4^-$ (Ba^{2+}, Cl^-) (Group I

(14) Add 6M HCl until the solution is acid. Test.

Al^{3+}, Cl^-, H^+ (Ba^{2+}) (Gro

(15) Add 3 drops aluminon reagent. Add 6M NH_3 until basic. Test.

$Al(OH)_3$
(R)

(Group IV)

My unknown contained the following cations ______________________________

You may wish to confirm your wet analysis by atomic absorption spectroscopy, described on page 342.

Notes

(11) a) The chromium ion may have formed the hydroxide complex in step 3. Hence the following steps are necessary to remove chromic ions from the procedure. This is done by oxidizing the ion to CrO_4^{2-} and then precipitating it with Ba^{2+}.

b) Both the chromium ion and aluminum ion show amphoteric properties in aqueous solutions. The hydroxides can react either with acids to form the metallic cation or with bases to form complex hydroxides.

$$Cr^{3+} \underset{H^+}{\overset{OH^-}{\rightleftharpoons}} Cr(OH)_3 \underset{H^+}{\overset{OH^-}{\rightleftharpoons}} Cr(OH)_4^-$$

$$Al^{3+} \underset{H^+}{\overset{OH^-}{\rightleftharpoons}} Al(OH)_3 \underset{H^+}{\overset{OH^-}{\rightleftharpoons}} Al(OH)_4^-$$

Aluminon reagent is adsorbed on the aluminum hydroxide producing a red gelatinous precipitate. This kind of adsorption phenomenon is called a lake.

(12) If this is a general unknown, substitute 0.2M $Pb(NO_3)_2$ for $BaCl_2$. A yellow precipitate of $PbCrO_4$ will form instead of $BaCrO_4$.

(14) Addition of HCl will precipitate $PbCl_2$ if $Pb(NO_3)_2$ was added in (12). *Keep solution cold* (ice). Discard precipitate. T.C.P.

(15) A lavendar precipitate will form if Pb^{2+} was present.

INVESTIGATION OF GROUP IV CATIONS

IDEAS

Group IV cations are precipitated as insoluble carbonates and the final identifications of each ion is done, not only by characteristic precipitate formation, but also by flame test.

A flame test is performed on the chloride salts because these salts have lower boiling points and hence are more volatile than the corresponding nitrates, sulfates, or oxalates. When heated in a bunsen flame, these cations will impart a specific color to the flame; for example, Ba^{2+} will color the flame light green. The heat of the flame excites the electrons to higher but unstable energy levels. The inevitable return to lower energy levels is accompanied by the emission of radiant energy of specific frequencies. Since the energies of electrons in each atom is unique, the amount of energy which can be absorbed or emitted is specific. The radiation is a "finger print" of the atom and makes identification possible.

INVESTIGATION

Chemicals: 0.1 M barium nitrate, $Ba(NO_3)_2$; 0.1M strontium nitrate, $Sr(NO_3)_2$; 0.1M calcium nitrate, $Ca(NO_3)_2$; 0.1M ammonium carbonate, $(NH_4)_2CO_3$; 1.0M potassium chromate, K_2CrO_4; 1M potassium oxalate, $K_2C_2O_4$; 1M ammonium sulfate, $(NH_4)_2SO_4$; 0.1M ammonium chloride, NH_4Cl 6M acetic acid, $HC_2H_3O_2$; 6M ammonia, NH_3; 6M sulfuric acid, H_2SO_4. Nichrome wire.

Perform each test in a 10-cm test tube using 2 drops of 0.1M cation solution diluted with 18 drops of water.

1. Prepare solutions of $Sr(NO_3)_2$, $Ca(NO_3)_2$, and $Ba(NO_3)_2$ in separate 10-cm test tubes. Add 5 drops 6M NH_3 and 5 drops 1M $(NH_4)_2CO_3$. Record your observations.

2. To a new set of diluted solutions of each cation add 5 drops 0.1M K_2CrO_4. Record your observations.

3. To a new set of the diluted cation solutions add 5 drops 1M$(NH_4)_2SO_4$. Heat. Record your observations.

4. To a new set of diluted cation solutions add 5 drops 1M $(NH_4)_2C_2O_4$. Record your observations.

5. Flame tests are done with a nichrome wire which has been cleaned by alternately dipping it into concentrated HCl and then burning off the acid in the hottest part of the bunsen flame until the flame remains colorless.

   Add 5 drops concentrated HCl to 2 drops 0.1M cation solution to be tested.*

---

* Usually these tests are done on the precipitates which have separated the cations from each other. Therefore, they are quite concentrated. If you do not obtain a clear test here, add several more drops of cation solution, evaporate the solution almost to dryness in a 30-ml beaker, and try the test again.

The clean nichrome wire is dipped in the cation solution and then put in the flame. The color is usually clearly visible for only a moment. Perform the test several times on each cation until you are sure of the identifying color. (A persistent yellow flame is due to sodium and indicates the wire has been contaminated by water, the desk surface, or your fingers!)

THOUGHT

Table IV

Characteristics of Group IV Cations

Give formulas of products as well as appearance.

Reagent	Ba^{2+}	Sr^{2+}	Ca^{2+}
1. 1M $(NH_4)_2CO_3$			
2. 1.0M K_2CrO_4			
3. 1.0M $(NH_4)_2SO_4$			
4. 0.1M $(NH_4)_2C_2O_4$			
5. Color of flame			

1. Using only one reagent, how would you separate the following pairs?

   a. Ba^{2+} from Sr^{2+}

   b. Sr^{2+} from Ca^{2+}

c. Ba^{2+} from Ca^{2+}

2. If you know that you have only one cation from Group IV, what is the simplest way to determine what it is?

GROUP IV CATIONS
FLOW CHART AND PROCEDURE

Solutions may contain 0.1M Ba^{2+}, 0.1M Sr^{2+}, 0.1M Ca^{2+} as NO_3^- which have been diluted 10 fold with deionized water. Use 2 ml.

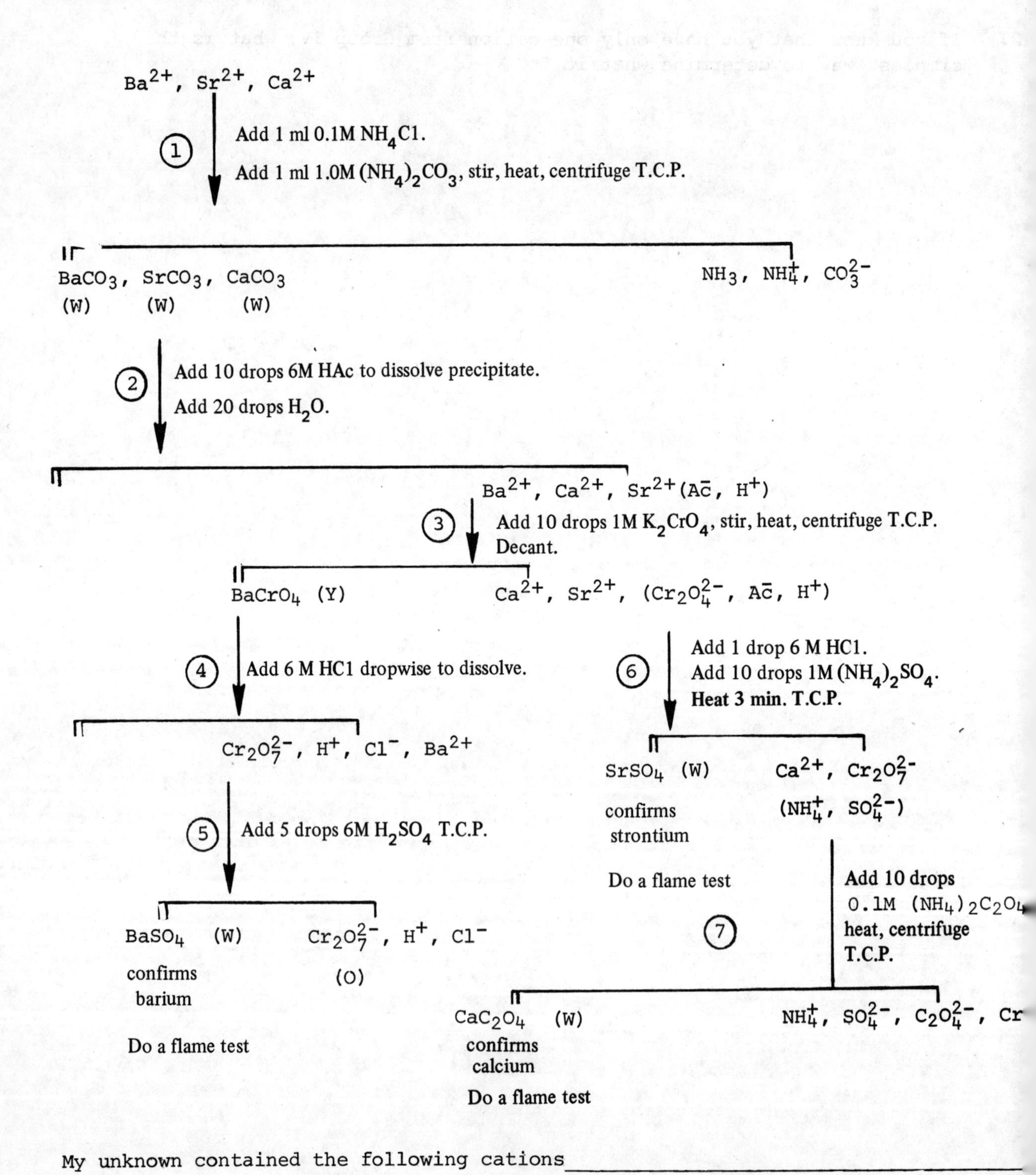

Ba^{2+}, Sr^{2+}, Ca^{2+}
1
Add 1 ml 0.1M NH_4Cl.
Add 1 ml 1.0M $(NH_4)_2CO_3$, stir, heat, centrifuge T.C.P.
$BaCO_3$, $SrCO_3$, $CaCO_3$
(W) (W) (W)
NH_3, NH_4^+, CO_3^{2-}
2
Add 10 drops 6M HAc to dissolve precipitate.
Add 20 drops H_2O.
Ba^{2+}, Ca^{2+}, Sr^{2+}($A\bar{c}$, H^+)
3
Add 10 drops 1M K_2CrO_4, stir, heat, centrifuge T.C.P.
Decant.
$BaCrO_4$ (Y)
Ca^{2+}, Sr^{2+}, ($Cr_2O_4^{2-}$, $A\bar{c}$, H^+)
4
Add 6 M HCl dropwise to dissolve.
$Cr_2O_7^{2-}$, H^+, Cl^-, Ba^{2+}
5
Add 5 drops 6M H_2SO_4 T.C.P.
$BaSO_4$ (W)
confirms barium
Do a flame test
$Cr_2O_7^{2-}$, H^+, Cl^-
(O)
6
Add 1 drop 6 M HCl.
Add 10 drops 1M $(NH_4)_2SO_4$.
Heat 3 min. T.C.P.
$SrSO_4$ (W)
confirms strontium
Do a flame test
Ca^{2+}, $Cr_2O_7^{2-}$
(NH_4^+, SO_4^{2-})
7
Add 10 drops
0.1M $(NH_4)_2C_2O_4$
heat, centrifuge
T.C.P.
CaC_2O_4 (W)
confirms calcium
Do a flame test
NH_4^+, SO_4^{2-}, $C_2O_4^{2-}$, Cr

My unknown contained the following cations ______________________

Notes. Group IV Cations

(2) The carbonates will dissolve in weakly acid solution with stirring and heat.

(3) $BaCrO_4$ will precipitate in a weakly acid solution. The solution should be slightly orange in color, indicating the presence of some $Cr_2O_7^{2-}$ ion.

$$\underset{(Y)}{2CrO_4^{2-}} + 2H^+ \longrightarrow \underset{(O)}{Cr_2O_7^{2-}} + H_2O$$

The solution must be slightly acidic to prevent the precipitation of $SrCrO_4$.

(6),(7) Procedure for flame test.

Obtain a nichrome wire and clean it thoroughly in concentrated hydrochloric acid. This is accomplished by repeatedly dipping the wire in the acid, and burning off the acid in a hot bunsen flame. The wire is clean when the flame's color is not changed by the acid on the wire.

Add 5 drops concentrated HCl (12M) to the precipitate to be tested, dip the clean platinum wire into the solution, then hold the wire in the flame.

Strontium gives a crimson-red color to the flame.
Calcium gives a brick-red color to the flame.

The only way to be sure of these flame identifications is to run the test on known strontium and calcium precipitates at the same time that the unknowns are done.

The purpose of the addition of concentrated hydrochloric acid to the sulfate and oxalate salts is to convert a minute quantity to the chloride. Chlorides are more volatile and give better flame tests than sulfates or oxalates.

A faint but persistenet yellow flame is due to the presence of sodium. The sodium ion is present in the deionized water and has also been introduced throughout the analysis.

NAME______________________________ SECTION____________ DATE__________

GROUP I CATIONS

The following are net ionic unbalanced equations.

Step 1. Add 6M HCl

1. $Ag^{+} + Cl^{-} \longrightarrow AgCl$

2. $Hg_2^{2+} + Cl^{-} \longrightarrow Hg_2Cl_2$

3. $Pb^{2+} + Cl^{-} \longrightarrow PbCl_2$

Step 2. Add hot water

4. $PbCl_2 \longrightarrow Pb^{2+} + Cl^{-}$

Step 3. Add 1M K_2CrO_4

5. $Pb^{2+} + CrO_4{}^{2-} \longrightarrow PbCrO_4$

Step 4. Add 6M NH_3 Confirm Mercury

6. $AgCl + NH_3 \longrightarrow Ag(NH_3)_2{}^{+} + Cl^{-}$

7. $Hg_2Cl_2 + NH_3 \longrightarrow HgNH_2Cl + NH_4{}^{+} + Cl^{-} + Hg^{\circ}$

Step 5. Add 6M HNO_3 Confirm silver

8. $Ag(NH_3)_2{}^{+} + H^{+} + Cl^{-} \longrightarrow AgCl + NH_4{}^{+}$

NAME______________________________ SECTION________________ DATE__________

GROUP II CATIONS

The following are net ionic unbalanced equations.

Step 1. Adjust pH to 0.5 and add H_2S.

1. $Cu^{2+} + H_2S \longrightarrow CuS + H^+$
2. $Hg^{2+} + H_2S \longrightarrow HgS + H^+$
3. $Cd^{2+} + H_2S \longrightarrow CdS + H^+$
4. $Sb^{3+} + H_2S \longrightarrow Sb_2S_3 + H^+$

Step 2. Add 6M NaOH and heat after washing above precipitates with H_2S in water. Analysis of these cations continues in Step 9.

5. $Sb_2S_3 + OH^- \longrightarrow Sb(OH)_4{}^- + S^{2-}$
6. $Sb_2S_3 + S^{2-} \longrightarrow SbS_2{}^-$
7. $[HgS + S^{2-} \longrightarrow HgS_2{}^{2-}]$

Step 3. Add 6M HNO_3

8. $CuS + H^+ + NO_3{}^- \longrightarrow Cu^{2+} + NO + S + H_2O$
9. $CdS + H + NO_3{}^- \longrightarrow Cd^{2+} + NO + S + H_2O$

Step 4. Add 6M HNO_3 and 6M HCl.

10. $HgS + NO_3{}^- + H^+ + Cl^- \longrightarrow HgCl_4{}^{2-} + NO + S + H_2O$

Step 5. Confirm mercury with copper penny.

11. $Hg^{+2} + Cu^\circ \longrightarrow Cu^{+2} + Hg^\circ$

Step 6. Add 6M NH_3 to confirm copper.

12. $Cu^{2+} + NH_3 \longrightarrow Cu(NH_3)_4{}^{2+}$ (blue solution)
13. $Cd^{2+} + NH_3 \longrightarrow Cd(NH_3)_4{}^{2+}$

Step 7. Add 6M NaOH to separate cadmium.

14. $Cd(NH_3)_4{}^{2+} + OH^- \longrightarrow$ Cd salt (Do not balance)

Step 8. Add 6M HCl, 6M NaOH to basic, H_2S to confirm cadmium.

15. Cd salt $+ H^+ + Cl^- \longrightarrow Cd^{2+}$ (Do not balance)
16. $Cd^{2+} + H_2S \longrightarrow CdS + H^+$

Step 9. Analysis of the solution from Step 2; add 12M HCl.

17. $Sb(OH)_4{}^- + H^+ + Cl^- \longrightarrow SbCl_6{}^{3-} + H_2O$

Step 10. Add water, H_2S to confirm antimony.

18. $SbCl_6{}^{3-} + H_2S \longrightarrow Sb_2S_3 + H^+ + Cl^-$

NAME________________________ SECTION____________ DATE____________

GROUP III CATIONS

Balance the following equations.

Step 1. Addition of 0.1M KSCN to confirm iron.

1. $Fe^{3+} + SCN^- \longrightarrow FeSCN^{2+}$

Step 2. Addition of 12 M HCl and heat. (This step is included for the general unknown, but omitted if only Group III cations are present.)

2. $H^+ + S^{2-} \longrightarrow HS^-$
3. $HS^- + H^+ \longrightarrow H_2S \uparrow$

Step 3. Addition of 6M NaOH and heat.

4. $Fe^{3+} + OH^- \longrightarrow Fe(OH)_3$
5. $Ni^{2+} + OH^- \longrightarrow Ni(OH)_2$
6. $Cr^{3+} + OH^- \longrightarrow Cr(OH)_3$
7. $[Cr^{3+} + OH^- \longrightarrow Cr(OH)_4^-]$
8. $Al^{3+} + OH^- \longrightarrow Al(OH)_4^-$

Step 4. To ppt. from Step 3, add 6M HCl.

9. $Fe(OH)_3 + H^+ \longrightarrow Fe^{3+} + H_2O$
10. $Ni(OH)_2 + H^+ \longrightarrow Ni^{2+} + H_2O$
11. $Cr(OH)_3 + H^+ \longrightarrow Cr^{3+} + H_2O$

Step 5. Add 6M NH_3.

12. $Fe^{3+} + NH_3 + H_2O \longrightarrow Fe(OH)_3 + NH_4^+$
13. $Cr^{3+} + NH_3 + H_2O \longrightarrow Cr(OH)_3 + NH_4^+$
14. $Ni^{2+} + NH_3 \longrightarrow Ni(NH_3)_6^{2+}$

Step 6. Add 1ml H_2O; Na_2O_2; heat.

15. $Cr^{3+} + Na_2O_2 + OH^- \longrightarrow CrO_4^{2-} + Na^+ + H_2O$

Step 7. Confirmation of nickel. Add dimethyl glyoxime.

16. $Ni(NH_3)_6^{2+} + C_4H_8N_2O_2 \longrightarrow Ni(C_4H_7N_2O_2)_2 + NH_4^+ + NH_3$

Step 8. Add 1M $BaCl_2$.

17. $Ba^{2+} + CrO_4^{2-} \longrightarrow BaCrO_4$
18. $BaCrO_4 + H^+ \longrightarrow Ba^{2+} + Cr_2O_7^{2-} + H_2O$

Step 10. Confirmation of chromium. Add H_2O_2.

19. $CrO_4^{2-} + H_2O_2 + H^+ \longrightarrow CrO_5 + H_2O$

Step 11. See equation for Step 6.

Step 12. See equation for Step 8.

Step 13. To Solution from Step 12, add 6M HCl.

20. $Al(OH)_4^- + H^+ \longrightarrow Al^{3+} + H_2O$

Step 14. Confirmation of aluminum. Add aluminon, 6M NH_3.

21. $Al^{+3} + NH_3 \cdot H_2O$ + aluminon $\longrightarrow Al(OH)_3$ red lake $+ NH_4^+$

NAME______________________________SECTION______________DATE______________

GROUP IV CATIONS

Balance the following equations.

Step 1. Add 0.1M NH_4Cl, + 1M$(NH_4)_2CO_3$. Heat.

1. $Ba^{2+} + CO_3^{2-} \longrightarrow BaCO_3$
2. $Sr^{2+} + CO_3^{2-} \longrightarrow SrCO_3$
3. $Ca^{2+} + CO_3^{2-} \longrightarrow CaCO_3$

Step 2. Add 6M HAc 1ml H_2O.

4. $BaCO_3 + H^+ \longrightarrow Ba^{2+} + H_2CO_3$
5. $SrCO_3 + H^+ \longrightarrow Sr^{2+} + H_2CO_3$
6. $CaCO_3 + H^+ \longrightarrow Ca^{2+} + H_2CO_3$

Step 3. Add 1M K_2CrO_4. Heat.

7. $Ba^{2+} + CrO_4^{2-} \longrightarrow BaCrO_4$
8. $CrO_4^{2-} + H^+ \longrightarrow Cr_2O_7^{2-} + H_2O$

Step 4. Add 6M HCl.

9. $BaCrO_4 + H^+ \longrightarrow Ba^{2+} + Cr_2O_7^{2-} + H_2O$

Step 5. Confirmation of barium. Add 6M H_2SO_4.

10. $Ba^{2+} + SO_4^{2-} \longrightarrow BaSO_4$

Step 6. Confirmation of strontium. Add 1M $(NH_4)_2SO_4$.

11. $Sr^{2+} + SO_4^{2-} \longrightarrow SrSO_4$

Step 7. Confirmation of calcium. Add 0.1M $Na_2C_2O_4$.

12. $Ca^{2+} + C_2O_4^{2-} \longrightarrow CaC_2O_4$

PROCEDURE FOR THE GENERAL UNKNOWN

The general unknown may contain as many as 14 cations and several anions. Begin the analysis with the cations using 3 ml of your unknown solution. Do not dilute this solution any further.

A general unknown differs from single group analysis in only 2 ways.

1. The first step in which the group is precipitated from solution requires that you retain both the precipitate and the solution, and that *all of the cations of a particular group be completely separated from succeeding groups* (i.e., T.C.P.) so that cations from earlier groups will not interfere with identifications later on in the analysis.

2. Not all cation groups may be represented in your unknown. For example, if your unknown contained Ag^{+}, Al^{3+}, and Ba^{2+}, there would be no Group II cations. This would be evident when no sulfide precipitates formed. In this case, the procedure is simply to be absolutely sure of the absence of the group (adding appropriate reagents, heating, etc.) and then going on to the next group.

The anion unknown is done on fresh samples of your unknown solution. Follow flow chart directions.

Sometimes, usually because of pH adjustments, the volume of solution exceeds the capacity of the 10-cm test tube. Evaporate the solution in your 50-ml beaker both to concentrate the ions and to reduce the volume. Then return the solution to the 10-cm tube and continue the analysis.

Usually the general unknown takes several laboratory periods, so you will have to interrupt your analysis and store your solutions. Keep a record on your flow chart showing clearly where you have stopped in the analysis. Close your test tubes with corks and label them to correspond with the steps in the flow chart.

QUALITATIVE ANALYSIS: ANIONS

IDEAS

The separation of anions into groups and their individual identification is not so straightforward as that for the cations. In cation analysis, the sequential isolation of the ions can be analytical because of the large differences in solubility of the soluble and insoluble salts. Many anions, on the other hand, form insoluble salts with more than one group reagent and precipitation within a particular group may not be complete. Complications also arise from the incompatibility of anions which can form oxidation-reduction pairs.

Anions generally encountered in inorganic analysis fall into one of five categories:

1. evolves as a gas when acidified with a strong acid
2. is an oxidizing agent in acid solution
3. is a reducing agent in acid solution
4. precipitates with $BaCl_2$ under slightly basic conditons
5. precipitates with $AgNO_3$ under neutral or acid conditions.

A selected group of eight anions have been chosen for analysis: Cl^-, Br^-, I^-, S^{-2}, SCN^-, SO_4^{-2}, PO_4^{-3} and NO_3^-.

All five categories are illustrated in this group with some anions falling into more than one category.

INVESTIGATION

Chemicals: 0.1M sodium chloride, NaCl; 0.1M sodium bromide, NaBr; 0.1M sodium iodide; 0.1M sodium thiocyannate, NaSCN; 0.1M sodium sulfide, Na_2S; 0.1M sodium sulfate, Na_2SO_4; 0.1M sodium phosphate, Na_3PO_4; 0.1M sodium nitrate, $NaNO_3$; 0.1M silver nitrate, $AgNO_3$; 6M ammonia, NH_3; 0.1M barium chloride, $BaCl_2$; 6M hydrochloric acid; 18M sulfuric acid, H_2SO_4; carbon tetrachloride, CCl_4; 15M nitric acid, HNO_3; 0.1M iron(III) chloride, $FeCl_3$; 1.5M sulfuric acid, H_2SO_4; 6M phosphoric acid, H_3PO_4; 0.5M ammonium molybdate, $(NH_4)_2MoO_4$; Iron(II) sulfate heptahydrate, $FeSO_4 \cdot 7H_2O$, solid. 1M barium acetate, $Ba(C_2H_3O_2)_2$ Lead acetate paper. Miller's reagent (see Note 6, page 340).

I. Rationale for Anion Analysis.

A. Separation into Anion Groups.

1. Set up 8 clean 10-cm test tubes in a rack. With *pencil*, write the formula for each anion on the frosted surface of each tube: Cl^-, Br^-. I^-, SCN^-, S^{-2}, SO_4^{-2}, PO_4^{-3}, NO_3^-. All solutions are 0.1M and are the sodium or potassium salts.

   To ½ ml of each solution, add 10 drops of 0.1M $AgNO_3$. Stir well and centrifuge. Note which solutions contain precipitates, what their colors are, and the nature of the solid (curdy or crystalline). Record observations in column 1 of Table IA.

   Clean the eight test tubes with deionized water and put in ½ ml of each anion solution as above.

   To each solution of the sodium salt, add 1 drop of 6M NH_3 and ½ ml of 0.1M $BaCl_2$.

   Note the formation and colors of all precipitates formed (T.C.P). Record your observations in column 2 of Table IA. Centrifuge, decant, and wash with deionized water any precipitates which formed.

   To those test tubes which contain a precipitate, add 4 to 5 drops of 6M HCl. Do any solids dissolve? Record your observations in column 3 of Table IA.

B. Determination of Compatibility of Anions

1. Effects of H_2SO_4 on Anion Stability.

Determine which anions can theoretically be oxidized by the presence of an oxidizing reagent.* Add ½ ml of each anion into separately labelled test tubes. Add 18 drops of deionized water and then 1 or 2 drops of concentrated (18M) H_2SO_4. Note whether any color changes occur, any gases escape, and carefully note any characteristic odor. Do not place your nose over the test tube, but hold the tube a few inches away and fan any gas toward your nose. Heat gently if no apparent reaction takes place. Cool and add 1 ml CCl_4, shake and note color of CCl_4 layer.

Record your observations in Table IB(1) on the data page and write balanced net ionic equations for any reactions which occur.

2. Effect of HNO_3 on Anion Stability.

Prepare solutions of the same group of anions as used in B1. To each solution add one or two drops of concentrated HNO_3(15M). Warm gently one minute, then cool, add 1 ml of CCl_4, shake and note color of CCl_4 layer. Record your observations in Table IB(2) on the data page. Write balanced net ionic equations for reactions which occur.

On the basis of these tests, determine which anions might not be compatible. The anion solutions you will study will not be as acidic as those in these tests. Thus, some reactions which you observed may not necessarily occur. However, if solutions of mixed anions are stored for a period of several weeks, there may be some deterioration in the composition. You are therefore to presume that any unknown you are given will not contain incompatible anions as determined by the tests in B1 and B2. This will make the analysis of the unknown much simpler since some initial positive tests for the presence of certain anions will rule out the presence of others. You should be aware of this in order to eliminate unnecessary testing.

List those anions which are incompatible in Table IB(3) on the data page. Have this list checked by your instructor.

C. Initial Individual Tests (to be performed on a small portion of the original unknown).

1. Spot test for Nitrate (NO_3^-).

Perform the spot test for NO_3^- on a known sample of NO_3^- and your unknown. Side by side comparison of known and unknown makes identification more reliable.

---

* In this series of tests, do not consider the oxygen (oxidation number -2) in any anion radical as one of the possibilities.

Brown Ring Test*.

Place a small crystal of $FeSO_4$ in a spot plate depression. Add 1 drop of solution to be tested and 1 drop of concentrated H_2SO_4. Formation of a brown ring around the crystal indicates the presence of $NO^-{}_3$

2. Test for the presence of SCN^-.

To 2 drops of unknown add 18 drops of H_2O in a 10-cm test tube. Add one or two drops of 0.1M Fe^{+++}. A blood red color will indicate the presence of SCN^-. In the presence of I^-, Fe^{+++} may impart a brownish yellow color but the $Fe(SCN)^{++}$ complex color is so intense and unmistakable that confusion as to the presence of SCN^- is unlikely.

3. Test for the presence of I^- and Br^-.

To 5 drops of unknown, add 5 drops of 1.5M H_2SO_4. Add 10 drops of 15M HNO_3 and 10 drops of CCl_4. Place in boiling water bath for *not more than 20 seconds*. Remove test tube from water bath and shake.

If I^- is present alone, a purple color will be imparted to the CCl_4 layer, due to the formation of I_2. If Br^- is present alone, the Br_2 formed will impart a brownish yellow color to the CCl_4 layer.

If both I^- and Br^- are present, the I_2 and Br_2 impart a murky deep brown color to the CCl_4 layer.

The intensity of the colors of the free halogens is enhanced in the organic layer (where they are more soluble than in the aqueous layer) making identification more accurate.

4. Test for the presence of S^{-2}.

To ½ ml of unknown in a 10-cm test tube, add a few drops of 6M HCl. Moisten a piece of lead acetate paper. Place the test tube in boiling water holding the wet paper over (not in) the mouth of the test tube. If S^{-2} is present, the paper will turn a dark brown due to the formation of PbS.

5. A test to distinguish between $BaSO_4$ and $Ba_3(PO_4)_2$.

Place 10 drops 1M $Ba(C_2H_3O_2)_2$ in each of two 10-cm test tubes. To one

* The reaction which takes place is:

$$3Fe^{2+} + NO_3^- + 4H^+ \longrightarrow 3Fe^{3+} + NO + 2H_2O$$

$$Fe^{2+} + NO \rightleftharpoons \underbrace{Fe(NO)^{2+}}_{\text{Brown Ring}}$$

Iodide can interfere with this test, but you should be able to determine on the basis of compatibility whether both NO_3^- and I^- will be found together in your unknown.

test tube add 5 drops 6M H_2SO_4; to the other add 5 drops 6M H_3PO_4. Stir each well, centrifuge, decant, and discard the solutions. Wash each precipitate with 5 drops water, centrifuge and discard the washings. Not the color of the precipitates. To each precipitate add 5 drops 6M HCl, and stir well. Record results on the data page.

6. Test for the presence of PO_4^{3-}.

   To 5 drops 6M H_3PO_4 in a 10-cm test tube, add 5 drops 15M HNO_3 (conc) and 5 drops 0.5M$(NH_4)_2MoO_4$. Warm in a water bath about one minute. A *yellow precipitate* of $(NH_4)_2PO_4 \cdot 12MoO_3$ (ammonium phosphomolybdate) forms slowly.

I. Rationale for Anion Analysis

Table IA

Separation into Anion Groups

Summary of anion reactions. Record color and formula of precipitates.

Anion	(1) 0.1M $AgNO_3$	(2) 0.1M $BaCl_2$ alkaline solution	(2) 0.1M $BaCl_2$ acid solution
Cl^-			
Br^-			
I^-			
SCN^-			
S^{2-}			
SO_4^{2-}			
PO_4^{3-}			
NO_3^-			

Table IB(1)

Determination of Compatibility of Anions

1. Effect of H_2SO_4 — Observations and formulas of products

Anions and 18M H_2SO_4	Product
Cl^-	
Br^-	
I^-	
SCN^-	
S^{2-}	
SO_4^{2-}	
PO_4^{3-}	
NO_3^-	

Table IB(2)

2. Effect of HNO_3 Observations and formulas of products.

Anions and 15 M HNO_3	Products
Cl	
Br^-	
I^-	
SCN^-	
S^{2-}	
SO_4^{2-}	
PO_4^{3-}	
NO_3^-	

3. Summary of compatibility

Table IB(3)

Anions Which Are Incompatible Because They React With Each Other

Cl^-	
Br^-	
I^-	
SCN^-	
S^{2-}	
SO_4^{2-}	
PO_4^{3-}	
NO_3^-	

Results of Individual Anion Tests IC

Give balanced net ionic equations which describe these reactions.

1. Brown Ring Test for NO_3^-

2. Test for SCN^-

3. Test for I^- and Br^- (Write two equations.)

4. Test for S^{2-}

5. Separation of $BaSO_4$ from $Ba_3(PO_4)_2$

	PO_4^{3-}	SO_4^{2-}
0.1M Ba^{2+}		
6M HCl		

6. Confirmation of PO_4^{3-}

7. Dissolving of AgCl with NH_3 (See Group I cation procedure.)

8. Reprecipitation of AgCl from the solution in (7). (See Group I cation procedure.)

THOUGHT

1. Use appropriate net ionic equations to illustrate the separation of the following anion pairs.

   a. Cl^- from NO_3^-

   b. SCN^- from SO_4^{2-}

c. SO_4^{2-} from PO_4^{3-}

2. Are there any anions which do not precipitate in the presence of Ag^+ or Ba^{2+}? Which are they, if any?

3. In the test for Br^- and I^- why is CCl_4 used?

4. Based on your experience with the anion tests, indicate to which category or categories outlined in the IDEAS section each anion belongs.

SEPARATION OF THE BARIUM GROUP FROM THE SILVER GROUP*
FLOW CHART AND PROCEDURE

Use 2 ml of the unknown anion solution.

Cl^-, Br^-, I^-, S^{2-}, SCN^-, NO_3^-, SO_4^{2-}, PO_4^{3-}

(1) Add 1M $Ba(C_2H_3O_2)_2$ dropwise until no further precipitation occurs. Stir, heat, centrifuge. (T.C.P.)

Precipitate: $BaSO_4$ (W), $Ba_3(PO_4)_2$ (W)

Solution: SCN^-, NO_3^-, $C_2H_3O_2^-$, Ba^{2+}, Cl^-, Br^-, I^-, S^{2-}

(continued p. 340 step (5))

(2) Wash ppt. with 10 drops of water. Discard washings. Add 5 drops 6M HC1. Stir and centrifuge. Decant. Save decantate.

Residue: $BaSO_4$

(3) Add 5 drops 6M HC1. Persistent white residue confirms the presence of sulfate ion.

Decantate: $H_2PO_4^-$

(4) Add 5 drops 15M HNO_3 and 5 drops 0.5M $(NH_4)_2MoO_4$. Warm one minute in water bath and let it stand.

Precipitate: $(NH_4)_2PO_4 \cdot 12MoO_3$ (Y) confirms PO_4^{3-}

Solution: NO_3^-, NH_4^+, H^+

Notes. Separation of the Barium Group from the Silver Group

(2) $Ba_3(PO_4)_2$ will dissolve in acid solution since PO_4^{-3} is the anion of a weak acid.

$$Ba_3(PO_4)_2 + 4H^+ \rightleftharpoons 3\ Ba^{++} + 2H_2PO_4^-$$

The $BaSO_4$ will not dissolve since H_2SO_4 is a strong acid and there is a negligible affinity of SO_4^{-2} for H^+.

(4) In strongly acid, warm solutions, ammonium molybdate reagent will give a yellow precipitate of ammonium phosphomolybdate.

$$H_2PO_4^- + 12MoO_4^{2-} + 22H^+ + 2NH_4^+ \longrightarrow (NH_4)_2\,PO_4 \cdot 12MoO_3 + 12H_2O$$

* For general unknown analysis, see page 329.

SEPARATION OF THE BARIUM GROUP FROM THE SILVER GROUP
FLOW CHART AND PROCEDURE (continued)

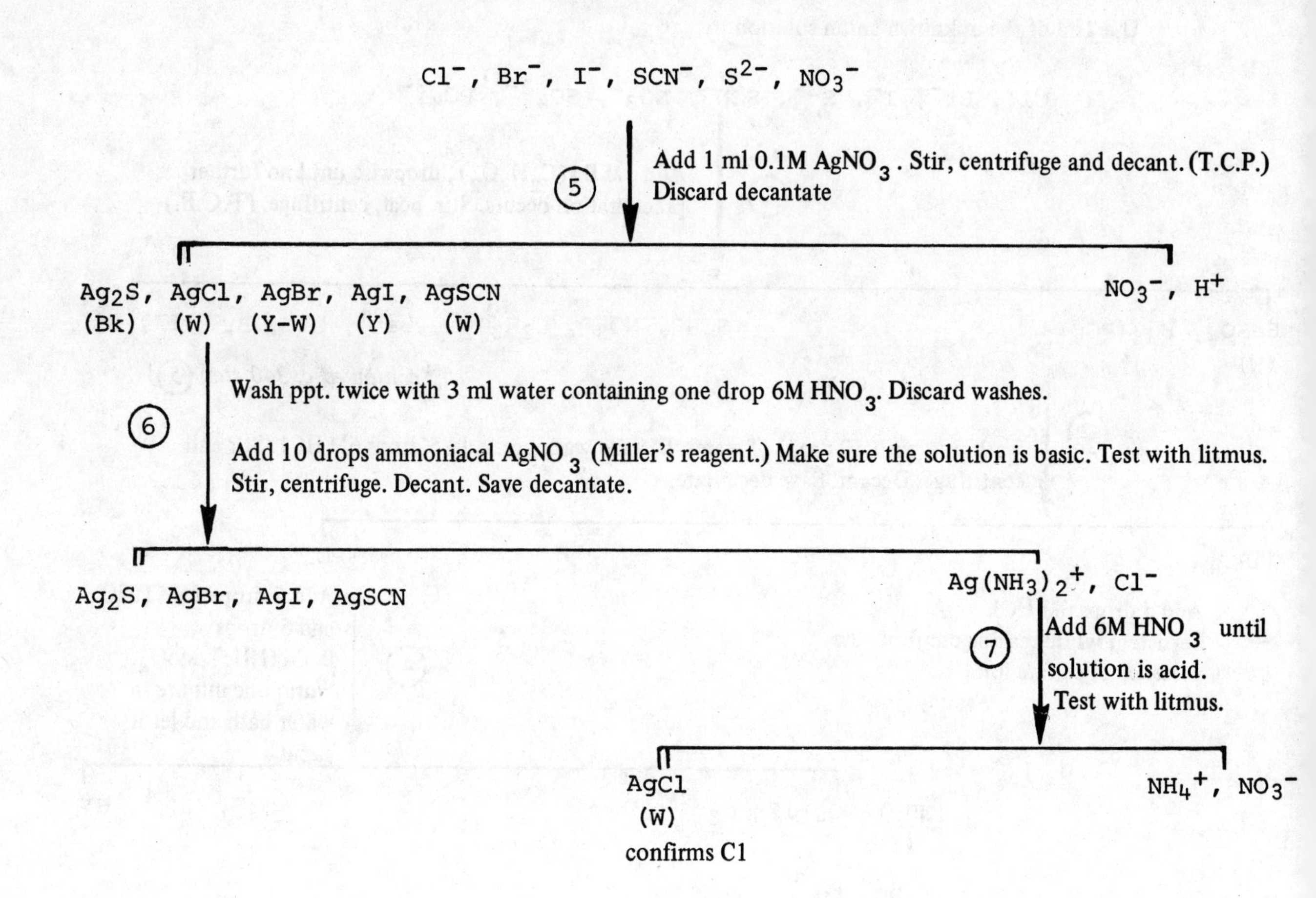

Cl^-, Br^-, I^-, SCN^-, S^{2-}, NO_3^-
5
Add 1 ml 0.1M $AgNO_3$. Stir, centrifuge and decant. (T.C.P.)
Discard decantate
Ag_2S, $AgCl$, $AgBr$, AgI, $AgSCN$
(Bk) (W) (Y-W) (Y) (W)
NO_3^-, H^+
6
Wash ppt. twice with 3 ml water containing one drop 6M HNO_3. Discard washes.
Add 10 drops ammoniacal $AgNO_3$ (Miller's reagent.) Make sure the solution is basic. Test with litmus.
Stir, centrifuge. Decant. Save decantate.
Ag_2S, $AgBr$, AgI, $AgSCN$
$Ag(NH_3)_2^+$, Cl^-
7
Add 6M HNO_3 until
solution is acid.
Test with litmus.
AgCl
(W)
confirms Cl
NH_4^+, NO_3^-

Notes

(5) If the preliminary test for S^{-2} was positive, then the precipitate at this point will be black. No specific test for sulfide is carried out at this point. The color of the silver salts and the result of the $Pb(C_2H_3O_2)_2$ paper test are sufficient to determine whether S^{-2} is present or not.

(6) AMMONIACAL SILVER NITRATE (Miller's Reagent)
1.7 g $AgNO_3$, 25g KNO_3, 17 ml 15M NH_3 dissolved in water and diluted to one liter in the presence of excess $Ag(NH_3)_2^+$ ions, AgI, AgBr, AgS are insoluble and AgSCN is only slightly soluble. AgCl will dissolve to form the more stable $Ag(NH_3)_2^+$.

$$AgCl + 2NH_3 \rightleftharpoons Ag(NH_3)_2^+ + Cl^-$$

(7) If Cl^- is present, this test should give a good precipitate. A slight cloudiness should be disregarded, especially if SCN^- has been confirmed by the test with Fe^{+++}.

ANIONS

NET IONIC EQUATIONS, UNBALANCED

IA. Summary of Anion Reactions

1. $Ag^+ + Cl^- \longrightarrow AgCl$
2. $Ag^+ + Br^- \longrightarrow AgBr$
3. $Ag^+ + I^- \longrightarrow AgI$
4. $Ag^+ + SCN^- \longrightarrow AgSCN$
5. $Ag^+ + S^{2-} \longrightarrow Ag_2S$
6. $Ag^+ + PO_4{}^{3-} \longrightarrow Ag_3PO_4$
7. $Ba^{2+} + PO_4{}^{3-} \longrightarrow Ba_3(PO_4)_2$ (alkaline solution)
8. $Ba^{2+} + SO_4{}^{2-} \longrightarrow BaSO_4$ (alkaline, or acid solution)
9. [$Ag^+ + SO_4{}^{2-} \longrightarrow Ag_2SO_4$]*

IB. Effect of Concentrate H_2SO_4 on Anion Stability

10. $I^- + H^+ + SO_4{}^{2-} \longrightarrow I_2 + SO_2 + H_2O$
11. $S^{2-} + H^+ + SO_4{}^{2-} \longrightarrow S^0 + SO_2 + H_2O$

Effect of Concentrate HNO_3 on Anion Stability

12. $S^{2-} + NO_3{}^- + H^+ \longrightarrow NO_2 + SO_3 + H_2O$
13. $I^- + H^+ + NO_3{}^- \longrightarrow NO_2 + H_2O + I_2$
14. $Br^- + H^+ + NO_3{}^- \longrightarrow NO_2 + H_2O + Br_2$
15. $SCN^- + H^+ + NO_3{}^- \longrightarrow SO_4{}^{2-} + NO + CO_2 + H_2O$

IC.

16. $Fe^{2+} + NO_3{}^- + H^+ \longrightarrow Fe^{3+} + NO + H_2O$
17. $Fe^{2+} + NO \longrightarrow Fe(NO)^{2+}$
18. $Fe^{3+} + SCN^- \longrightarrow Fe(SCN)^{2+}$
19. $I^- + H^+ + NO_3{}^- \longrightarrow I_2 + NO + H_2O$
20. $Br^- + H^+ + NO_3{}^- \longrightarrow Br_2 + NO + H_2O$
21. $S^{2-} + H^+ \longrightarrow H_2S$
22. $H_2S + Pb(C_2H_3O_2)_2 \longrightarrow PbS + C_2H_3O_2{}^- + H^+$
23. $BaSO_4 + H^+ \longrightarrow$ No Reaction
24. $Ba_3(PO_4)_2 + H^+ \longrightarrow Ba^{2+} + H_2PO_4{}^-$
25. $PO_4{}^{3-} + MoO_4{}^{2-} + H^+ + NH_4{}^+ \longrightarrow (NH_4)_3PO_4 \cdot 12MoO_3 + H_2O$
26. $AgCl + NH_3 \longrightarrow Ag(NH_3)_2^+ + Cl^-$
27. $Ag(NH_3)_2^+ + H^+ + Cl^- \longrightarrow AgCl + NH_4{}^+$

---

* Square brackets indicate that this reaction may occur under certain conditions.

QUALITATIVE ANALYSIS: INSTRUMENTAL METHOD

ATOMIC ABSORPTION

Absorption and radiation occur when a valence electron of an atom is raised to a higher energy state and then returns to the lower energy level. This phenomemon is utilized in atomic absorption. Atomic absorption has been an invaluable tool in analytical determinations. It is simple, quick, accurate, and sensitive to very low concentrations of cations.

The essential features of an atomic absorption instrument are:

a) A radiation source: a hollow cathode tube which is designed to generate radiation of the specific frequency required to promote electrons to a higher energy level. The emitted radiation is characteristic of the metal within the hollow cathode.

b) A means of convertings the cations to atoms: a reducing flame (acetylene) converts the aqueous ionic solution into droplets, evaporates these to a hydrated metal residue, dehydrates the salt, and then reduces the cations to atoms.

c) A detection device to measure the attenuation (decrease in signal streng produced when the radiation from the cathode tube is passed through the sample.

In applying atomic absorption to qualitative investigations, it is only necessary to determine that a particular cation is present or absent. Suitable concentrations of cation solutions for use with this instrument contain about 10 micrograms of cation per milliliter of solution.

Dilute one milliliter of the unknown solution to 100 milliliters to get an appropriate concentration. The specific mode of operation of the instrument available to you will be demonstrated by your instructor.

QUALITATIVE ANALYSIS: PAPER CHROMATOGRAPHY

IDEAS

Chromatography means "written in color" in Greek. It is a technique used for the separation of different materials on an adsorbent. Separation occurs because of the varying degrees of affinity that different substances exhibit for a particular adsorbent, the presence and position of the particular substances being indicated by their characteristic colors. The word "chromatography" was coined by M. Tswett (1872-1920) whose name coincidentally also means "color" in Russian. Tswett,* a botanist, and pioneering biochemist, first used a column of adsorbent to separate chlorophyll pigments of plants. He described his method of column chromatography in his first paper on the subject in 1906:

"If a petroleum ether chlorophyll solution filters through a column of adsorbent (I use chiefly calcium carbonate which is tightly tamped into a glass tube), the pigments separate themselves from top to bottom in different colored zones according to the adsorption series, in which the more strongly adsorbed substances displace the more weakly held ones farther toward the bottom. This separation is practically complete if one, after the passage of the pigment solution, passes a stream of pure solvent through the column of adsorbent. Just as the rays of light in the spectrum, so are the different components of a pigment mixture separated in an orderly way in the calcium carbonate column, and they may be qualitatively and also quantitatively determined in this way. Such a preparation I call a chromatogram and the corresponding method, the chromatographic method. Needless to say, the described adsorption phenomena are suited not only to the chlorophyll pigments. It is to be expected that all kinds of colored or colorless chemical compounds are subject to the same laws."**

The technique of paper chromatography, in which porous filter paper was used instead of an adsorbent column, was developed in 1944 by A. Martin and R. Synge. With this technique, it was possible to separate amino acids in protein molecules, and other complex substances and mixtures. For instance, after a drop of amino acid has been placed near the bottom of a paper strip and allowed to dry, the strip is dipped into a solvent. Because of capillary action, the solvent travels slowly up the strip, carrying with it the components of the dried amino acids. The rate of flow depends upon the solubility of the particular amino acids in the solvent and in water, and so the components are separated. The presence and location on the strip of the various components may be detected by chemical reaction with an appropriate reagent, and the components may be identified by comparing with a strip on which known substances have diffused. Martin and Synge received the Nobel Prize in chemistry in 1952 as the innovators of the technique of paper chromatography.

---

* He was truly international. He was born in Italy of a Russian father and Italian mother and lived in Switzerland, Poland, Estonia, and Russia.

** M. Tswett, Ber. deutch, bot. Ges., 24, 316-323 (1906) quoted in *Chymia*, Vol. 6, H. Leicester, ed., University of Pennsylvania Press, Philadelphia, 1960, p. 150.

In 1953, Martin developed the new technique of gas chromatography. The material to be separated is evaporated and this vapor, usually with an inert carrier gas, travels through a column in which the different components have different speeds.

INSTRUCTIONS FOR THE ANALYSIS OF GROUP II AND GROUP III CATIONS

In this experiment , paper is used as the adsorbent and the cations of Groups II and III are the materials which will exhibit their unique affinities for the paper.

The technique is carried out quite simply by placing a spot of the cation solution to be analyzed near the end of a strip of paper. The paper strip is hung vertically so the end just below the cation spot is immersed in an inorganic solvent in which polarity has been augmented by hydrochloric acid. Due to capillary action, the solvent rises slowly up the paper, washing over the cation spot. If the cation has great affinity for the paper, the spot will not be carried by the rising solvent. However, if the cation is not strongly adsorbed by the paper, it will be carried along by the moving solvent.

I. Discussion of the Separation and Identification of Cations

A. Colored Spots

1. While the chromatograms are running and the atmosphere in the closed container is saturated with the solvent, some cations develop colors due to the formation of complexes with the solvent. Observe which cations do this. Note also that the color usually disappears when the paper is removed from the solvent environment.

2. Most of the cations will be colorless on the paper. In order to locate the positions of these cations after the chromatogram is run, it is necessary to spray them with a reagent which will cause the formation of identifying spots.

   Group II: All Group II cations produce colored sulfides. Therefore, after the chromatogram has been run, the paper is sprayed with ammonium sulfide to locate the cations.

   Group III: After the chromatogram has been run, the paper is sprayed with 8-hydroxyquinoline. This reagent forms complexes with the cations which fluoresce or absorb ultra violet light.

B. The Rf value

Each cation rises a unique distance in a given solvent mixture. Therefore this distance is also used to identify the cation. To eliminate the differences which a single cation would move due to variations in the length of time that the chromatogram is run, the distances are recorded as a ratio, called R_f. As soon as the chromatogram run is discontinued the solvent front is marked with a pencil line. When the cation spot is developed, two distances are measured: one is from the base line to center of the cation spot.

$$\frac{\text{distance the cation moved}}{\text{distance the solvent moved}} = R_f$$

The R_f value for a cation can then be compared with a known value which had been determined on the same kind of paper under similar conditions. Often knowns and unknowns are run simultaneously.

The R_f value is never greater than 1.

Chemicals:

Solvent A: 90 ml acetone, 5 ml concentrate hydrochloric acid, and 4 ml water.
Solvent B: 45 ml ethanol, 45 ml methanol, 60 ml 2M hydrochloric acid.
6M hydrochloric acid, HCl.

Spray solutions:
To develop colored sulfide spots: 1M ammonium sulfide, $(NH_4)_2S$
To develop fluorescent or absorbing spots, combine reagents in the following order:
2 g 8-hydroxyquinoline
20 ml ethanol
Saturated sodium acetate ($NaC_2H_3O_2$), a few drops
Glacial acetic acid, $HC_2H_3O_2$, dropwise with stirring until the color of the solution becomes a clear yellow

Equipment:

Strips of chromatography paper, 21 cm long
100-ml graduated cylinders with corks to fit
1 50-ml beaker
1 dropper
2 sprayers (either aerosol or hand pump)
ultra violet light

Procedure:

The purpose of the following procedure is to convert the nitrates, which were introduced with both the copper nitrate and cadmium nitrate, to chlorides. Chlorides seem to allow more uniform movement of the cations up the paper. Because some of the chlorides have very low boiling points,* care must be taken during the evaporating process not to heat the solution to dryness.

Put about 2 ml of your Group II or Group III unknown in a clean small beaker. Add 10 drops 6M HCl and evaporate until the volume of solution is reduced to about ½ ml. You may notice some brown fumes. These come from the decomposition of the nitrates which were used to make two of the unknown solutions. Repeat the addition of 10 drops of 6M HCl and evaporate again to ½ ml. *Avoid heating to dryness.* Some chlorides can be lost by boiling away.

The solutions are concentrated so that a small spot will provide a detectable quantity of cation.

---

* Boiling points of some chlorides: $SbCl_3$, 283°C; $CuCl_2$, 993°C; $HgCl_2$, 302°C; $CdCl_2$, 960°C.

The chromatograms for Groups II and III may be run simultaneously in different graduated cylinders. For each, the procedure is the same; only different solvents are used.

For each chromatogram, cut a strip of paper 21 cm long. If you are setting up the standards for comparison, Group II would require a chromatogram for each cation (4) and one for your unknown. Group III would also require one for each cation (4) and one for the unknown. With a *pencil*, draw a line across each strip about 2.5 cm from one end (bottom). Put the formula for the cation or the Group number at the other end (top). Then with a clean dropper, after discarding the first drop, place one small spot of unknown on the pencil line at the bottom of the strip. The drop must *not* spread out to the edges of the paper. A convenient technique is to place the dropper on the paper and then gently squeeze the bulb. *Allow the spot to dry.*

For each strip, obtain a 100-ml graduated cylinder with a cork to fit the cylinder, and a paper clip. Straighten the longer loop of each clip. Push this wire into the smaller end of the cork until the remaining loop of the clip is up against the cork. Attach the top end of the paper strip to the clip so that the paper is not punctured. The cation spot is on the end furthest from the cork.

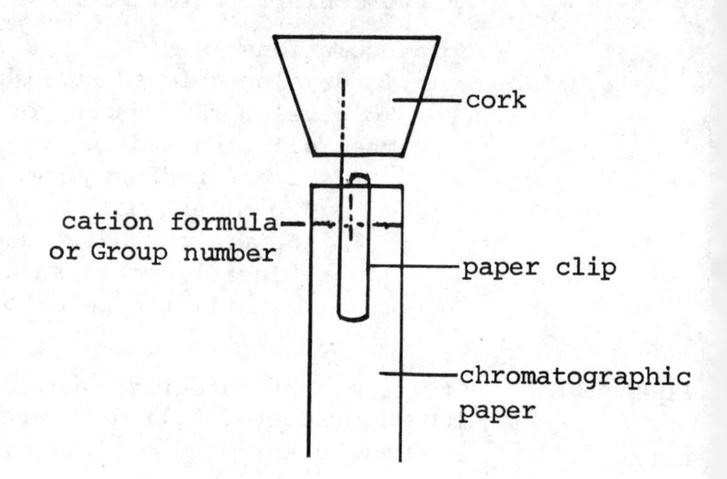

Group II Cations : Put about 8 ml solvent mixture B in a 100 ml graduated cylinder. Hang the paper strip in the cylinder so that the end just reaches the solvent but the cation spot is *above* the solution level. (The length of the strip may require adjustment. It should hang just below the liquid level but not touch the bottom of the cylinder or the sides when the cork is in place.) Insert the cork securely in the cylinder.

Group III Cations : Put about 8 ml solvent mixture A in a graduated cylinder and follow the instructions given above.

While the solvent front is advancing, observe the paper to detect any colored spots. Although these spots may disappear when the paper is removed from the solvent atmosphere, this is an important clue to the identification of the cation.

The experiment is stopped when the solvent front is just below the paper clip. (Paper clips although usually covered with a film of plastic, are attacked by the solvent. The metal cations from the clip will react with developers and obscure the identification of cations from either the unknown or from the standards which may have cations that have kept pace with the moving solvent front.) The rate of flow of the solvent is affected by ambient temperatures and is faster when the air is warm. However, for solvent mixture A the time is about one hour, and for solvent mixture B the time is about 3 hours.

Remove the paper from the cylinder. Mark the solvent front with a pencil, and allow the strip(s) to air dry for about 15 minutes. Place the strip(s) on separate sheets of clean paper and attach them with scotch tape at the end which had been attached to the clip. Holding the sheet vertically in the hood, gently spray the Group II strip with ammonium sulfide, and the Group III strip with 8-hydroxyquinoline. Do not direct a strong spray at the strip, because the spray may drive the cation through the strip behind it.

For Group II cations, observe the spots and record the color and R_f value of each on the table on the data page. Compare your results with the standard and identify the cations.

For Group III cations, use ultra violet light to view the strip. You will notice that prolonged exposure to the radiation will make the strip fluoresce or absorb more light. The spots will also be enhanced if they are exposed to ammonia fumes. This can be done by passing the strips *over* a small beaker containing concentrated ammonia on the hood. While the strips are under the ultra violet light, outline the spots with pencil and indicate with an A whether the spot is dark and absorbing light, or with an F to indicate that the spot is bright and is fluorescing.

NAME____________________SECTION__________DATE__________

Chromatographic Analysis

Group II

Cation	Color spots due to solvent	Formula of sulfide	Distance solvent traveled (cm)	Distance cation traveled	R_f
Cu^{2+}					
Hg^{2+}					
Cd^{2+}					
Sb^{3+}					
Unknown					

Group II unknown is

Group III

Cation	Color spots due to solvent	Absorbed light	Fluoresced	Distance solvent traveled (cm)	Distance cation traveled (cm)	Rf
Al^{3+}						
Fe^{3+}						
Ni^{2+}						
Cr^{3+}						
Unknown						

Group III unknown is

APPENDIX A. CONVERSION FACTORS

LENGTH

Unit	Equivalent Values
1 meter	6.214×10^{-4} mile
	3.281 ft
	39.37 inches
	1.000×10^{-3} km
	100.0 cm
	1000 mm
	1.000×10^{10} Å
	1.000×10^{6} μ (micron)
	1.000×10^{9} mμ (millimicron) or nanometers

VOLUME

Unit	Equivalent Values
1 liter	1.000×10^{3} cm^3
	1.000×10^{3} ml
	0.2642 gal
	1.057 qt
	6.102×10^{4} cu inches

MASS

Unit	Equivalent Values
1 kilogram	2.205 lb
	1.000×10^3 g

TEMPERATURE

Unit	Equivalent Values
0° Celsius	32°F
100° Celsius	212°F
-273.16° Celsius	0°K

OTHER USEFUL EQUIVALENTS

Unit	Equivalent Values
1 atomic mass unit	1.660×10^{-24} g
1 atmosphere pressure	1.013×10^5 newtons/m^2
	14.70 lbf/in.
	760 mm Hg or torr

SOME CONSTANTS

Avogadro's number, N	6.023×10^{23} molecules/mole
Universal gas constant, R	8.205×10^{-2} liter atm/(mole)(deg K)
	1.987 cal/(mole)(deg K)
	8.314 kJ/(mole) (deg K)
Mechanical equivalent of heat, J	4.185 joules/cal
British thermal unit (B.T.U.)	252 cal

APPENDIX B. MASS AND WEIGHT

Mass and weight, often used interchangeably, are related but not identical concepts. Mass is a measure of the resistance of an object to a force; weight refers to a particular force: the gravitational force between the earth and the object. If a mass is caused to accelerate, the concept "force" is evoked as the cause of the acceleration. Force is measured as the product of mass times acceleration. Mathematically expressed, F = ma. The gram is an arbitrarily chosen standard of mass. If 1 gram accelerates 1 cm per second per second, then the force is defined as 1 dyne. If 1 gram accelerates 980 cm/sec^2, then the force is 980 dynes. Since 980 cm/sec^2 is the approximate **gravitational acceleration of all objects at the surface of the earth, then the weight of a mass of 1 gram is 980 dynes.**

Although we speak of "weighing" something on a platform balance, the units of mass are used. This is actually inconsistent; it would be preferable to say that we are "massing" something. Unfortunately, the verb **"massing"** is nonexistent. In the operation of weighing on the platform balance, a known weight is balanced against an unknown, or one can alternatively assert that the force which the earth exerts on the known is equal to the force on the unknown. Since the acceleration of gravity is the same for all objects at any given location, acceleration cancels out, and the masses of the objects are actually compared.

Expressed mathematically, this becomes

F_g = weight of an object

a_g = acceleration of gravity = 980 cm/sec^2

$F_g = ma_g$

At the equilibrium position of the balance:

unknown F_g = known F_g

unknown ma_g = known ma_g

APPENDIX C. IUPAC NOMENCLATURE

In 1959 the International Union of Pure and Applied Chemistry agreed on a system of nomenclature for inorganic compounds that assigns a unique name to each possible compound. However, older systems of nomenclature are still being used, as well as trivial and common names.

BINARY COMPOUNDS

Containing only two types of elements, these compounds are named using the suffix "ide": for example, magnesium nitride, Mg_3N_2; sodium chloride, NaCl. If the electropositive element has more than one oxidation state, a roman numeral in parentheses is used following the name of the cation: for example, iron (III) fluoride, FeF_3; chlorine (VII) oxide, Cl_2O_7; nitrogen (II) oxide, NO.

When a particular element of a given oxidation state occurs in different compounds, Greek prefixes are used to denote the varying numbers of atoms: nitrogen dioxide, NO_2; dinitrogen tetraoxide, N_2O_4; phosphorus pentachloride, PCl_5; tetraphosphorus hexaoxide, P_4O_6.

SALTS OF POLYPROTIC ACIDS

Greek prefixes are also used here: for example, disodium monohydrogen phosphate, Na_2HPO_4.*

SALTS OF OXYGEN ACIDS

For the names of acids that end with the suffix "ic," the suffix "ate" is used for the salt: for example, a salt of nitric acid (HNO_3) is potassium nitrate, KNO_3. The suffix "ite" is used for the salts of the "ous" acids: for example, a salt of sulfurous acid (H_2SO_3) is potassium hydrogen sulfite, $KHSO_3$.

COMPLEX CATIONS

The following sequence is used in naming compounds of complex cations:

1. Greek prefixes to specify the number of ligands. Ligands are negative ions, negatively charged groups, or neutral molecules surrounding a central cation
2. The name of the ligand usually ends in "o"; one exception is ammine, NH_3.

* The prefix "mono" may be omitted when there is no ambiguity in the naming.

The central metal ion followed by a roman numeral in parentheses indicating its oxidation state.

Some examples are tetraaquocopper (II) sulfate, $Cu(H_2O)_4SO_4$; diammine silver (I) chloride, $Ag(NH_3)_2Cl$. If both neutral and negative ligands are contained in the complex ion, the negative ligands are named first: for example, bromopentammine cobalt (II) sulfate, $CoBr(NH_3)_5SO_4$.

COMPLEX ANIONS

These are named according to the followng sequence:

1. The cation with parenthetical roman numeral to specify the oxidation state

2. The Greek prefix designating the number of ligands

3. Name of the ligand

4. The name of the central metal ion using the Latin root ending with the suffix "ate"

5. A roman numeral in parentheses giving the oxidation state of the metal.

   Some examples are silver (I) hexacyanoferrate (II), $Ag_4Fe(CN)_6$; sodium tetrahydroxodiaquoaluminate, $NaAL(OH)_4(H_2O)_2$; potassium hexacynoferrate (III), $K_3Fe(CN)_6$.

APPENDIX D.
SPECIFICATIONS OF SOME CONCENTRATED ACIDS AND AMMONIUM HYDROXIDE

Substance	Molecular weight	Molarity	Percent solute[a]	Specific gravity
Acetic acid (glacial), $HC_2H_3O_2$	60.05	17	99-100	1.05
Hydrochloric acid, HCl	36.46	12	36.5-38	1.19
Nitric acid, HNO_3	63.01	16	69.6	1.42
Sulfuric acid, H_2SO_4	98.08	18	95-98	1.84
Phosphoric acid, H_3PO_4	98.00	15	85-86	1.7
Ammonium hydroxide, $NH_3 \cdot H_2O$ (aqueous ammonia)	35.05	15	28-30 as NH_3 58-62 as $NH_3 \cdot H_2O$	0.90

[a] Slight variations are attributable to different manufacturers and different grades.

APPENDIX E.
EQUILIBRIUM CONSTANTS AND SOLUBILITY PRODUCTS

IONIZATION CONSTANTS OF WEAK ACIDS

Acetic acid	$HC_2H_3O_2 + H_2O = H_3O^+ + C_2H_3O_2^-$	$K = 1.76 \times 10^{-5}$
Carbonic acid	$H_2CO_3 + H_2O = H_3O^+ + HCO_3^-$	$K_1 = 4.3 \times 10^{-7}$
	$HCO_3^- + H_2O = H_3O^+ + CO_3^{2-}$	$K_2 = 5.61 \times 10^{-11}$
Oxalic acid	$H_2C_2O_4 + H_2O = H_3O^+ + HC_2O_4^-$	$K_1 = 5.9 \times 10^{-2}$
	$HC_2O_4^- + H_2O = H_3O^+ + C_2O_4^{2-}$	$K_2 = 6.4 \times 10^{-5}$
Phosphoric acid	$H_3PO_4 + H_2O = H_3O^+ + H_2PO_4^-$	$K_1 = 7.52 \times 10^{-3}$
	$H_2PO_4^- + H_2O = H_3O^+ + HPO_4^{2-}$	$K_2 = 6.23 \times 10^{-8}$
	$HPO_4^{2-} + H_2O = H_3O^+ + PO_4^{3-}$	$K_3 = 2.2 \times 10^{-13}$
Acid sulfate	$HSO_4^- + H_2O = H_3O^+ + SO_4^{2-}$	$K_2 = 1.2 \times 10^{-2}$
Sulfurous acid	$H_2SO_3 + H_2O = H_3O^+ + HSO_3^-$	$K_1 = 1.54 \times 10^{-2}$
	$HSO_3^- + H_2O = H_3O^+ + SO_3^{2-}$	$K_2 = 1.02 \times 10^{-7}$
Hydrosulfuric acid	$H_2S + H_2O = H_3O^+ + HS^-$	$K_1 = 1.1 \times 10^{-7}$
	$HS^- + H_2O = H_3O^+ + S^{2-}$	$K_2 = 1.0 \times 10^{-15}$

IONIZATION CONSTANT OF A WEAK BASE

Ammonium hydroxide	$NH_3 \cdot H_2O = NH_4^+ + OH^-$	$K_1 = 1.79 \times 10^{-5}$

Strong acids and strong bases are those which are soluble in water and which have a primary ionization that is virtually 100 percent. A few common examples are

hydrochloric acid
nitric acid
sulfuric acid
sodium hydroxide
potassium hydroxide
barium hydroxide

SOLUBILITY PRODUCTS

Acetates:

Silver acetate	$AgC_2H_3O_2 = Ag^+ + C_2H_3O_2^-$	$K_{sp} = 2.5 \times 10^{-3}$

Carbonates:

Barium carbonate	$BaCO_3 = Ba^{2+} + CO_3^{2-}$	$K_{sp} = 8.1 \times 10^{-9}$
Calcium carbonate	$CaCO_3 = Ca^{2+} + CO_3^{2-}$	$K_{sp} = 1.2 \times 10^{-8}$
Magnesium carbonte	$MgCO_3 = Mg^{2+} + CO_3^{2-}$	$K_{sp} = 2.0 \times 10^{-4}$
Silver carbonate	$Ag_2CO_3 = 2Ag^+ + CO_3^{2-}$	$K_{sp} = 8 \times 10^{-12}$
Strontium carbonate	$SrCO_3 = Sr^{2+} + CO_3^{2-}$	$K_{sp} = 1.6 \times 10^{-9}$

Chromates:

Barium chromate	$BaCrO_4 = Ba^{2+} + CrO_4^{2-}$	$K_{sp} = 2 \times 10^{-10}$
Lead (II) chromate	$PbCrO_4 = Pb^{2+} + CrO_4^{2-}$	$K_{sp} = 1.8 \times 10^{-14}$
Silver chromate	$Ag_2CrO_4 = 2Ag^+ + CrO_4^{2-}$	$K_{sp} = 9 \times 10^{-12}$

Halides:

Lead chloride	$PbCl_2 = Pb^{2+} + 2Cl^-$	$K_{sp} = 1 \times 10^{-4}$
Lead iodide	$PbI_2 = Pb^{2+} + 2I^-$	$K_{sp} = 1.4 \times 10^{-8}$
Mercury (I) chloride	$Hg_2Cl_2 = 2Hg + 2Cl^-$	$K_{sp} = 2 \times 10^{-19}$
Silver chloride	$AgCl = Ag^+ + Cl^-$	$K_{sp} = 1.2 \times 10^{-10}$
Silver bromide	$AgBr = Ag^+ + Br^-$	$K_{sp} = 4.1 \times 10^{-13}$
Silver iodide	$AgI = Ag^+ + I^-$	$K_{sp} = 1.0 \times 10^{-16}$

Hydroxides:

Aluminum hydroxide	$Al(OH)_3 = Al^{3+} + 3OH^-$	$K_{sp} = 4 \times 10^{-13}$
Calcium hydroxide	$Ca(OH)_2 = Ca^{2+} + 2OH^-$	$K_{sp} = 5.8 \times 10^{-6}$
Chromium (III) hydroxide	$Cr(OH)_3 = Cr^{3+} + 3OH^-$	$K_{sp} = 6 \times 10^{-31}$
Copper (II) hydroxide	$Cu(OH)_2 = Cu^{2+} + 2OH^-$	$K_{sp} = 5.6 \times 10^{-20}$
Iron (II) hydroxide	$Fe(OH)_2 = Fe^{2+} + 2OH^-$	$K_{sp} = 1.6 \times 10^{-14}$
Iron (III) hydroxide	$Fe(OH)_3 = Fe^{3+} + 3OH^-$	$K_{sp} = 1.1 \times 10^{-36}$

Magnesium hydroxide	$Mg(OH)_2 = Mg^{2+} + 2OH^-$	$K_{sp} = 1.2 \times 10^{-11}$
Nickel hydroxide	$Ni(OH)_2 = Ni^{2+} + 2OH^-$	$K_{sp} = 1 \times 10^{-15}$
Zinc hydroxide	$Zn(OH)_2 = Zn^{2+} + 2OH^-$	$K_{sp} = 4.5 \times 10^{-7}$
Oxalates:		
Barium oxalate	$BaC_2O_4 = Ba^{2+} + C_2O_4{}^{2-}$	$K_{sp} = 1.2 \times 10^{-7}$
Calcium oxalate	$CaC_2O_4 = Ca^{2+} + C_2O_4{}^{2-}$	$K_{sp} = 2.3 \times 10^{-9}$
Sulfates:		
Barium sulfate	$BaSO_4 = Ba^{2+} + SO_4{}^{2-}$	$K_{sp} = 1 \times 10^{-10}$
Calcium	$CaSO_4 = Ca^{2+} + SO_4{}^{2-}$	$K_{sp} = 2.4 \times 10^{-5}$
Lead sulfate	$PbSO_4 = Pb^{2+} + SO_4{}^{2-}$	$K_{sp} = 1 \times 10^{-8}$
Strontium sulfate	$S_rSO_4 = S_r{}^{2+} + SO_4^{2-}$	$K_{sp} = 2.8 \times 10^{-7}$
Sulfides:		
Cadmium sulfide	$CdS = Cd^{2+} + S^{2-}$	$K_{sp} = 3.6 \times 10^{-29}$
Copper (II) sulfide	$CuS = Cu^{2+} + S^{2-}$	$K_{sp} = 4 \times 10^{-38}$
Iron (II) sulfide	$FeS = Fe^{2+} + S^{2-}$	$K_{sp} = 3.7 \times 10^{-19}$
Iron (III) sulfide	$Fe_2S_3 = 2Fe^{3+} + 3S^{2-}$	$K_{sp} = 10^{-88}$ (?)
Lead (II) sulfide	$PbS = Pb^{2+} + S^{2-}$	$K_{sp} = 3.4 \times 10^{-28}$
Mercury (II) sulfide	$HgS = Hg^{2+} + S^{2-}$	$K_{sp} = 4 \times 10^{-53}$
Silver sulfide	$Ag_2S = 2Ag^+ + S^{2-}$	$K_{sp} = 1 \times 10^{-50}$
Zinc sulfide	$ZnS = Zn^{2+} + S^{2-}$	$K_{sp} = 1.2 \times 10^{-23}$

APPENDIX F. SOLUBILITY TABLE

Explanation of Table

Solubility is given in grams per 100 ml at temperatures t_1 and t_2. The first recorded solubility at t_1, includes the range from 15° to 25°C, except where noted. The second, at t_2, is at 100° C except where noted.

s = soluble; vs = very soluble; sl s = slightly soluble; v sl s = very slightly soluble; i = insoluble; d = decomposes; hydr = hydrolyzes.

Colors: w = white; r = red; o = orange; y = yellow; g = green; bl = blue; v = violet; br = brown; blk = black; p = pink; col = colorless.

Key

molecular weight color water of crystallization, if any solubility t_1 solubility t_2

In general, these solubility rules apply to the ions mentioned in this laboratory manual: all nitrates are soluble; the hydroxides and salts of ammonium, potassium, and sodium are soluble.

The source of the information in this table is the *Handbook of Physics and Chemistry*, R. C. Weast ed., The Chemical Rubber Co., Cleveland, Ohio, 50th ed., 1969-1970.

	Ag^+	Al^{3+}	Ba^{2+}	Ca^{2+}	Co^{2+}	Cu^{2+}
Br^-	187.8 y - 8.4×10^{-6} 3.7×10^{-4}	266.7 w - s -	333.19 w $2H_2O$ 60 151	199.9 w - 142 -	326.8 r $6H_2O$ s 153	233.3 bl - vs -
$C_2H_3O_2^-$	166.9 w - 1.02 2.52(80°)	204.1 w - s but hydr -	273.45 w H_2O 43.3 75	176.2 w H_2O 37.4(0°) 29.7	249.1 r $4H_2O$ s s	199.65 g H_2O 78.5 -
CO_3^{2-}	275.8 w - 3.2×10^{-3} 5×10^{-2}		197.35 w - 2×10^{-3} 6×10^{-3}	100.1 w - 1.45×10^{-3} -	118.9 r - i i	
$C_2O_4^{2-}$	303.8 w - 4×10^{-3} -	390.1 w $4H_2O$ i -	225.4 w - 9.3×10^{-3} -	128.1 w - 7×10^{-4} -	182.98 p $2H_2O$ v sl s sl s	160.6 bl $1/2H_2O$ 2.53×10^{-3} -
Cl^-	143.3 w - 1.5×10^{-4} 2.1×10^{-3}	133.3 w - 69.9 -	244.3 w $2H_2O$ 26.3 37.0	110.99 w - 74.5 -	237.9 r $6H_2O$ 76.7 76.7	134.4 y - 70.6 107.9

SOLUBILITY TABLE

	Ag^+	Al^{3+}	Ba^{2+}	Ca^{2+}	Co^{2+}	Cu^{2+}
$CrO_4{}^{2-}$	331.73 r - 2.5×10^{-3} 8×10^{-3} (70°)		253.3 y - 3.8×10^{-4} -	192.1 y $2H_2O$ 16.3 -	174.9 gray - i d	
I^-	234.8 y - 2.5×10^{-7} 2.5×10^{-6} (60°)	407.7 br - sd -	391.15 w - 66.5 -	293.89 y - 2.09 -	420.8 r $6H_2O$ s s	
OH^-		78.00 w - i i	315.48 col $8H_2O$ 5.6 94.7(78°)	74.09 col - 0.185(°) 0.077	92.95 r - 3.2×10^{-1} -	97.56 bl - i d
$PO_4{}^{3-}$	418.6 y - 6 10^{-} -	121.95 w - i -	601.96 w - i i	310.2 w - 2.5×10^{-3} -	402.8 p $2H_2O$ i -	434.6 bl $3H_2O$ i i
S^{2-}	247.8 blk - 2×10^{-5} v sl s	150.2 y - d	169.4 w - d -	72.14 w - 2.1×10^{-2} d		95.60 blk - 3.3×10^{-5} -
SCN^-	165.95 w - 2×10^{-5} 6.4×10^{-4}		289.5 w $2H_2O$ 4.3 -	210.3 w $3H_2O$ vs vs	229.1 v $3H_2O$ s -	121.7 blk - d d
$SO_3{}^{2-}$	295.8 w - sl s -		217.4 w - 2×10^{-2} 2×10^{-3}(80°)	156.2 w $2H_2O$ 4.3×10^{-3} 1.7×10^{-1}	229.1 r $5H_2O$ i -	
$SO_4{}^{2-}$	311.8 w - 8 10^{-} 1.41	342.15 w - 31.3(0°) 98.1	233.4 w - 2.4×10^{-4} 4.1×10^{-4}	136.14 w - 2.08×10^{-1} 1.6×10^{-1}	263.1 r $6H_2O$ - -	249.7 bl $5H_2O$ 31.6 203.3

SOLUBILITY TABLE

	Fe^{2+}	Fe^{3+}	Mg^{2+}	Ni^{2+}	Pb^{2+}	Zn^{2+}
Br^-	215.7 gy – 109(10°) 170(95°)	295.6 r – s –	184.1 w – 101.5 125.6	272.6 g $3H_2O$ 199(0°) 316	367.01 w – 4.55×10^{-1}(0°) 4.71	225.2 w – 447 675
$C_2H_3O_2^-$	246.0 g $4H_2O$ vs –		142.4 w – vs vs	248.9 g $4H_2O$ – –	325.3 w – 44.3 221(50°)	183.5 w – 30 44.6
CO_3^{2-}	115.85 gray – 6.7×10^{-3} –		84.3 w – 1.06×10^{-2} –	118.7 g – 9.3×10^{-3} i	267.2 w – 1.1×10^{-4} d	125.4 w – 1.0×10^{-3} –
$C_2O_4^{2-}$	179.9 y $2H_2O$ 2.2×10^{-2} 2.6×10^{-2}	465.8 y $5H_2O$ vs –	148.4 w $2H_2O$ 7.0×10^{-2} 8.0×10^{-2}	182.8 g $2H_2O$ i –	295.2 w – 1.6×10^{-4} –	189.4 w $2H_2O$ 7.9×10^{-4} –
Cl^-	198.8 gy $4H_2O$ 160.1(10°) 415.5	162.2 br – 74.4(0°) 535.7	95.22 w – 54.25 72.7	237.7 g $6H_2O$ 254 599	278.1 w – 9.9×10^{-1} 3.34 0.99*	136.3 w – 432 615
CrO_4^{2-}			266.4 y $7H_2O$ vs –		323.2 y – 5.8×10^{-6} i	233.4 y – i d
I^-	309.7 gray – s –		278.1 w – 148 164.9(110°)	312.5 blk – 124 188	461.0 y – 4.4×10^{-2}(0°) 4.1×10^{-1}	319.2 w – 432 511
OH^-	89.87 g – 6.7×10^{-4} –	106.86 br – i i	58.33 col – 9×10^{-4} 4×10^{-3}		241.2 w – 1.55×10^{-2} sl s	99.38 w – v sl s –

* at 0°C

SOLUBILITY TABLE

	Fe^{2+}	Fe^{3+}	Mg^{2+}	Ni^{2+}	Pb^{2+}	Zn^{2+}
PO_4^{3-}	501.6 bl $8H_2O$ i -	186.85 y $2H_2O$ v sl s 6.7×10^{-1}	262.9 irid. - i i	510.2 g $8H_2O$ i i	811.5 w - 1.4×10^{-5} -	386.05 w - i i
S^{2-}	87.9 blk - 6.2×10^{-4} d	207.9 yg - v sl s d	56.38 r br - d d	90.77 blk - 3.6×10^{-4} -	239.25 blk - 1.24×10^{-2} -	97.43 w - 6.9×10^{-4} -
SCN^-	226.1 g $3H_2O$ vs -	230.1 r - vs d			323.35 w - 5×10^{-2} 2×10^{-1}	181.5 w - 2.6 8
SO_3^{2-}	189.96 y $3H_2O$ v sl s -		212.5 w $6H_2O$ 66 s	246.9 g $6H_2O$ i -	287.25 w - i i	181.5 w $2H_2O$ 1.6×10^{-1} d
SO_4^{2-}	278.05 g $7H_2O$ 15.65 48.6(50°)	399.9 y - sls d	120.4 w - 26(0°) 73.8	262.86 g $6H_2O$ 62.5 340.7	303.24 w - 4.25×10^{-3} 5.6×10^{-3}	161.4 w - - -

APPENDIX G
STANDARD OXIDATION-REDUCTION POTENTIALS *

Half-reaction	$E°$ (volts)
$Li^+ + e \rightarrow Li$	-3.045
$K^+ + e \rightarrow K$	-2.924
$Ba^{2+} + 2e \rightarrow Ba$	-2.90
$Ca^{2+} + 2e \rightarrow Ca$	-2.76
$Na^+ + e \rightarrow Na$	-2.7109
$PO_4{}^{3-} + 2H_2O + 2e \rightarrow HPO_3{}^{2-} + 3OH^-$	-1.05
$2H_2O + 2e \rightarrow H_2 + 2OH^-$	-0.8277
$Zn^{2+} + 2e \rightarrow Zn$	-0.7628
$Ag_2S + 2e \rightarrow 2Ag + S^{2-}$	-0.7051
$S + 2e \rightarrow S^{2-}$	-0.508
$2CO_2 + 2H^+ + 2e \rightarrow H_2C_2O_4$	-0.49
$Fe^{2+} + 2e \rightarrow Fe$	-0.409
$2H^+ + 2e \rightarrow H_2$	0.0000
$NO_3{}^- + H_2O + 2e \rightarrow NO_2{}^- + 2OH^-$	0.01
$AgSCN + e \rightarrow Ag + SCN^-$	0.0895
$SO_4{}^{2-} + 4H^+ + 2e \rightarrow H_2SO_3 + H_2O$	0.20
$AgCl + e \rightarrow Ag + Cl^-$	0.2223
$Hg_2Cl_2 + 2e \rightarrow 2Hg + 2Cl^-$	0.2682
$Cu^{2+} + 2e \rightarrow Cu$	0.3402
$O_2 + 2H_2O + 4e \rightarrow 4OH^-$	0.401

Half-reaction (cont'd)	$\mathcal{E}$° (volts) (cont'd)
$Ag_2CrO_4 + 2e \rightarrow 2Ag + CrO_4^{2-}$	0.4463
$Fe^{3+} + e \rightarrow Fe^{2+}$ (0.5M H_2SO_4)	0.679
$Fe(phenanthroline)_3^{3+} + e \rightarrow Fe(ph)_3^{2+}$ (2 M H_2SO_4)	1.056
$Cr_2O_7^{2-} + 14H^+ + 6e \rightarrow 2Cr^{3+} + 7H_2O$	1.33
$MnO_4^- + 8H^+ + 5e \rightarrow Mn^{2+} + 4H_2O$	1.491
$1/2F_2 + H^+ + e \rightarrow HF$	3.03

*All species are at unit activity and at 25°C.

The source of the values in this table is the Handbook of Physics and Chemistry, R. C. Weast, ed., The Chemical Rubber Co., Cleveland, Ohio, 50th ed., 1969-1970.

$\mathcal{E}$° is voltage of the standard half-cell compared to the standard hydrogen electrode. The standard hydrogen electrode is arbitrarily assigned the value of zero.

APPENDIX H. ATOMIC WEIGHT

(Atomic Weights Based on Carbon 12)

Name	Symbol	Atomic weight[a]
Actinium	Ac	(227)
Aluminum	Al	26.9815
Americium	Am	(243)
Antimony	Sb	121.75
Argon	Ar	39.948
Arsenic	As	74.9216
Astatine	At	(210)
Barium	Ba	137.34
Berkelium	Bk	(247)
Beryllium	Be	9.0122
Bismuth	Bi	208.980
Boron	B	10.811
Bromine	Br	79.909
Cadmium	Cd	112.40
Calcium	Ca	40.08
Californium	Cf	(249)
Carbon	C	12.01115
Cerium	Ce	140.12
Cesium	Cs	132.905
Chlorine	Cl	35.453
Chromium	Cr	51.996
Cobalt	Co	58.9332
Copper	Cu	63.54
Curium	Cm	(247)
Dysprosium	Dy	162.50
Einsteinium	Es	(254)
Erbium	Er	167.26
Europium	Eu	151.96
Fermium	Fm	(253)
Fluorine	F	18.9984
Francium	Fr	(223)
Gadolinium	Gd	157.25
Gallium	Ga	69.72
Germanium	Ge	72.59
Gold	Au	196.967
Hafnium	Hf	178.49
Helium	He	4.0026
Holmium	Ho	164.930
Hydrogen	H	1.00797

Name	Symbol	Atomic Weight[a]
Indium	In	114.82
Iodine	I	126.9044
Iridium	Ir	192.2
Iron	Fe	55.847
Krypton	Kr	83.80
Lanthanum	La	138.91
Lawrencium	Lw	(257)
Lead	Pb	207.19
Lithium	Li	6.939
Lutetium	Lu	174.97
Magnesium	Mg	24.312
Manganese	Mn	54.9380
Mendelevium	Md	(256)
Mercury	Hg	200.59
Molybdenum	Mo	95.94
Neodymium	Nd	144.24
Neon	Ne	20.183
Neptunium	Np	(237)
Nickel	Ni	58.71
Niobium	Nb	92.906
Nitrogen	N	14.0067
Nobelium	No	(254)
Osmium	Os	190.2
Oxygen	O	15.9994
Palladium	Pd	106.4
Phosphorus	P	30.9738
Platinum	Pt	195.09
Plutonium	Pu	(244)
Polonium	Po	(210)
Potassium	K	39.102
Praseodymium	Pr	140.907
Promethium	Pm	(145)
Protactinium	Pa	(231)
Radium	Ra	(226)
Radon	Rn	(222)
Rhenium	Re	186.2
Rhodium	Rh	102.905
Rubidium	Rb	85.47
Ruthenium	Ru	101.07
Samarium	Sm	150.35
Scandium	Sc	44.956
Selenium	Se	78.96
Silicon	Si	28.086
Silver	Ag	107.870
Sodium	Na	22.9898
Strontium	Sr	87.62
Sulfur	S	32.064
Tantalum	Ta	180.948
Technetium	Tc	(99)
Tellurium	Te	127.60
Terbium	Tb	158.924
Thallium	Tl	204.37

Name	Symbol	Atomic Weight[a]
Thorium	Th	232.038
Thulium	Tm	168.934
Tin	Sn	118.69
Titanium	Ti	47.90
Tungsten	W	183.85
Uranium	U	238.03
Vanadium	V	50.942
Xenon	Xe	131.30
Ytterbium	Yb	173.04
Yttrium	Y	88.905
Zinc	Zn	65.37
Zirconium	Zr	91.22

[a] Values in parentheses are mass numbers of isotopes of longest half-life.

APPENDIX I.
PERIODIC CHART
(Electron Distribution of the Elements)

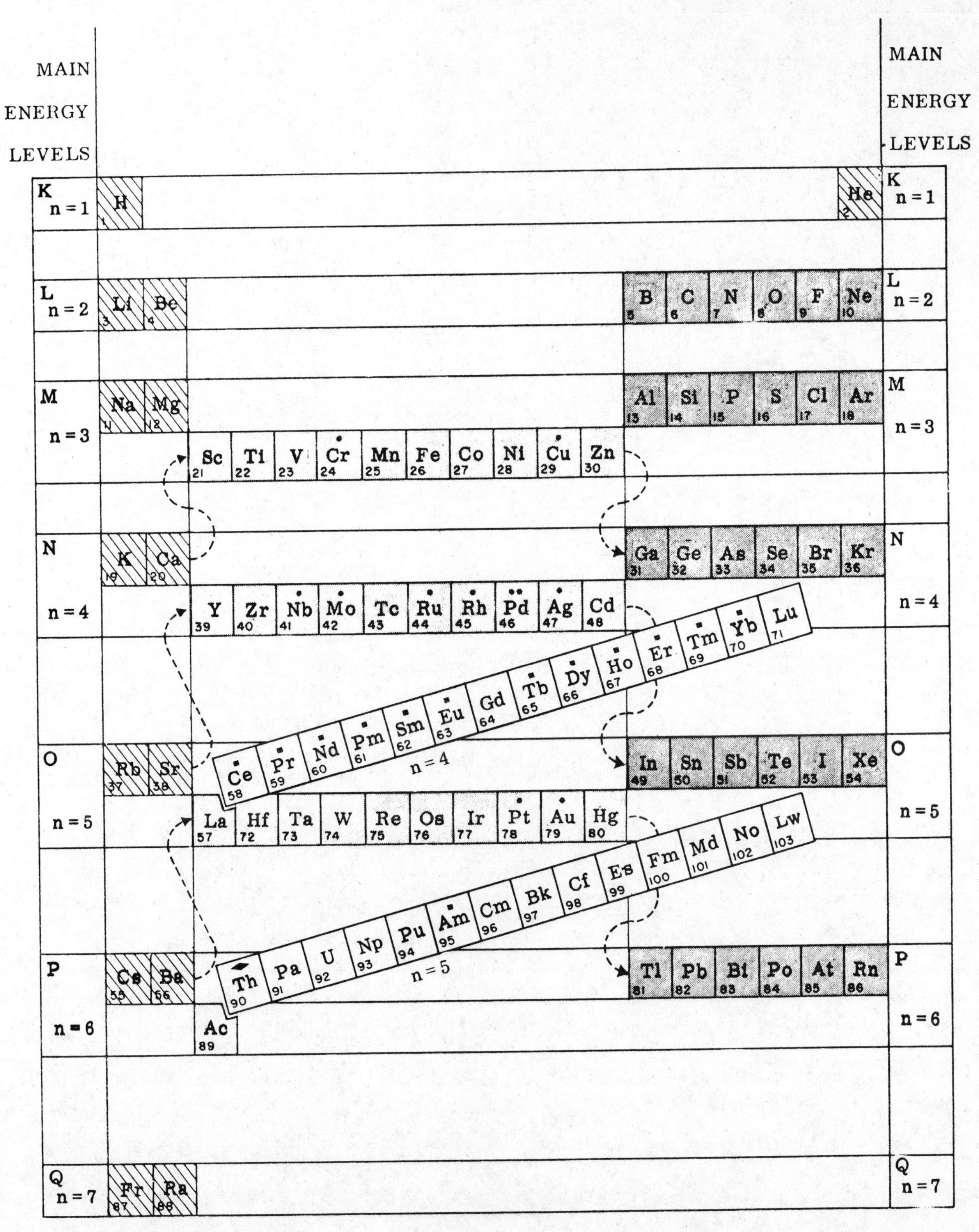

"s" orbitals, ℓ=0 ▧
"p" orbitals, ℓ=1 ■
"d" orbitals, ℓ=2 □
"f" orbitals, ℓ=3 □

• an outer s electron occupies a "d" orbital
▪ an outer d electron occupies an "f" orbital
◆ the expected f electron occupies a "6d" orbital